Climate Change

Climate Change

IPCC, Water Crisis, and Policy Riddles with Reference to India and Her Surroundings

Binayak Ray

LEXINGTON BOOKS
Lanham • Boulder • New York • Toronto • Plymouth, UK

Published by Lexington Books
A wholly owned subsidiary of The Rowman & Littlefield Publishing Group, Inc.
4501 Forbes Boulevard, Suite 200, Lanham, Maryland 20706
www.lexingtonbooks.com

Estover Road, Plymouth PL6 7PY, United Kingdom

British Library Cataloguing in Publication Information Available

Library of Congress Cataloging-in-Publication Data

Ray, Binayak.
Climate change : IPCC, water crisis, and policy riddles with reference to India and her surroundings / Binayak Ray.
p. cm.
Includes bibliographical references and index.
ISBN 978-0-7391-6414-3 (cloth : alk. paper) — ISBN 978-0-7391-6863-9 (electronic)
1. Climatic changes—India. 2. Climatology—India. 3. Fresh water—India. 4. Water-supply—India. 5. Environmental policy—India. 6. India—Environmental conditions. I. Title.
QC903.2.I4R39 2011
363.738'740954—dc23 2011023795

∞™ The paper used in this publication meets the minimum requirements of American National Standard for Information Sciences—Permanence of Paper for Printed Library Materials, ANSI/NISO Z39.48-1992.

Printed in the United States of America

I dedicate this book to the Honourable Tom Uren, former minister, Department of Urban and Regional Development, Department of Territories and Local Government, Commonwealth of Australia.

I also dedicate this book to my uncle, Dr. Dilip Kumar Pal, formerly of the United Nations Development Programme, and to my teacher Dr. Subrata Gupta.

Contents

Tables

Abbreviations

ABC	Australian Broadcasting Corporation
ADB	Asian Development Bank
AWG	Ad Hoc Working Group
BNP	Bangladesh Nationalist Party
CENTO	Central Treaty Organization
CIMMYT	International Maize and Wheat Improvement Center
CITES	Convention on International Trade in Endangered Species
CoP	Conference of Parties
CSIRO	Commonwealth Scientific and Industrial Research Organisation
EBRD	European Bank for Reconstruction and Development
EEA	European Environment Agency
EU	European Union
FAO	Food and Agriculture Organization
FEER	Far Eastern Economic Review
GHG	Greenhouse Gas
ICAO	International Civil Aviation Organization
ICJ	International Court of Justice
IFRC	International Federation of Red Cross and Red Crescent Societies
IFPRI	International Food Policy Research Institute
IMF	International Monetary Fund
IPCC	Intergovernmental Panel on Climate Change
IPR	Intellectual Property Right
IRRI	International Rice Research Institute
ISI	Inter-services Intelligence unit (Pakistan)
MRC	Mekong River Commission

NAAC	North Atlantic Cooperation Council
NAM	Non-aligned Movement
NASA	National Aeronautics and Space Administration
NATO	North Atlantic Treaty Organization
PRC	Peoples' Republic of China
SAARC	South Asian Association for Regional Cooperation
SEATO	South East Asia Treaty Organization
TRIPS	Trade Related Aspects of Intellectual Property Rights
UN	United Nations
UNCSD	United Nations Commission for Sustainable Development
UNDESA	United Nations Department of Economic and Social Affairs
UNEP	United Nations Environment Program
UNFCCC	United Nations Framework Convention on Climate Change
UNWWDR	United Nations World Water Development Report
WFP	World Food Program
WHO	World Health Organization
WMO	World Meteorological Organization
WTO	World Trade Organization

Note: Unless otherwise mentioned, all currencies referred to in the book are in US dollars.

Preface

The Bali Climate conference in 2007 is a watershed in the annals of the climate change debate for three reasons. First, those skeptical countries that had refused to ratify the Kyoto Protocol earlier (such as Australia under the conservative government of John Howard), signed the Protocol (in Australia's case, the liberal government of Kevin Rudd had assumed power). Second, although the US government under George W. Bush was staunchly against the Kyoto Protocol, it did not oppose the Bali Declaration that sought to establish a new protocol in two years' time. Third, the mood of the conference was so intense that even those countries that had been lukewarm or hostile toward the Kyoto Protocol softened their positions and indicated that something had to be done to save the planet from the ravages of a rapidly deteriorating global environment.

Scientists worldwide have been arguing for some time that while precise prediction of the impacts of climate change may be impossible, the rapid deterioration of global climate has reached a point of no return. The effects of the rise in global temperature (and its impacts on rainfall, availability of fresh water, food production, human health, and national security) are far-reaching.

Freshwater and unpolluted air—two of the world's most precious natural resources—are becoming scarcer and are reaching beyond the bounds of a large section of the world's poverty-stricken populations (globally, one person in five lives below the poverty line). Aside from the humanitarian consequences of these catastrophic developments, climate change also threatens global security.[1] There is broad agreement worldwide that human activities are responsible for the current situation. Some within the scientific community believe that the speed of deterioration is so fast that if urgent corrective steps are not taken, we may reach a point of no return.

The climatic and water environment in South Asia has been deteriorating for some time and if governments and communities do not adopt appropriate policies on these issues as a priority, countries in this region will face unprecedented catastrophe. Most rivers in northeast India straddle country borders. Water and water-related policy responses are required at a regional level, and unfortunately this has not been the case so far. Europe also has many such rivers, yet it has managed to pursue a cooperative policy even when the Cold War was at its height. The effects of climate change have disastrous consequences for all countries in South Asia—a region already embroiled in numerous internal and external conflicts.

Some of the reasons why South Asian countries are failing to pursue an effective regional policy are identified and explored in this book. European countries recognize the necessity for integrated climate change and water resource management policies. Experience has taught them that a fragmented policy is ineffective. Over the years they have gradually shifted their thinking toward cooperative transboundary water management of water resources and response to climate change issues.[2] The South Asian Association for Regional Cooperation has existed for decades; yet the political leadership has been unable to achieve institutional arrangements for climate change or transboundary water management. The political structures in a society determine the parameters within which decision making occurs,[3] and the absence of a regional approach in South Asia reflects the political reality.

Sociopolitical issues are complex in South Asia, where mutual distrust and hostility are common between countries. Even at the best of times these countries find it difficult to resolve their differences, even though failure to address the issues of climate change and water crisis would be disastrous for everyone. This book identifies the critical issues and suggests that it lies with political leadership of countries in the region to decide on a course of action to overcome the consequences of climate change and water crisis; no outsider can undertake this task for them.

NOTES

1. Barlow 2007: 145.
2. Louka 2008: 401.
3. Chapman 2000: 252.

Acknowledgments

I sincerely thank my colleagues for their encouragement. Special thanks go to Emeritus Fellow Dr. Ronald J. May, without whose support it would have been difficult to complete the book. Thanks also go to Professor Paul Hutchcroft and Dr. Ed Aspinall for providing departmental support. I also thank Mr. Eric Johns for his help with the computer work, Ms. Allison Ley for her editorial assistance, and Mr. Nur Ahmed for his assistance with the research work.

I also wish to acknowledge the help provided by the reference librarian and staff of the Petherick Reading Room at the National Library of Australia. This book would not have been completed without their assistance.

Introduction

Rapid deterioration of the global climate and the speed at which it is changing has become the centerpiece of all public policies in the new millennium. The consequences of these changes and the collective human inability to live within the limited supply of natural resources (including freshwater) have raised the fundamental question of sustainability. While at various levels the climate change and sustainability debate has gradually crept into the global policy debate, it is only in the late twentieth century that the gravity of the crises has dawned on the collective human consciousness.

The global scientific community mostly agrees that human activities are primarily responsible for this deterioration; but a small but powerful section refuses to accept this position. Their counterargument has been that the changes are due to nature's own cyclical phenomenon. Adherents to this belief are taking full advantage of the fact that climatology as a scientific discipline is in its infancy. The issues are far more complex than one would like to believe. Scientists are gradually overcoming the knowledge gap, and have already provided evidence that human behavior and activities are responsible for our rapidly deteriorating environment.

Impacts from climate change have extensive ramifications and cover every aspect of human civilization. It is mostly accepted that the root cause of climate change is the unrestricted emission of greenhouse gas (or CO_2) into the atmosphere. And the burning of fossil fuel is the primary source of greenhouse gas. Scientists agree that this gas is causing the Earth's surface temperature to rise, and that failure to control CO_2 would be disastrous for humanity. Most vulnerable in a catastrophe are the economically and politically marginalized sections of a community.

Global collective efforts during the end of the twentieth century led to an agreement known as the Kyoto Protocol, which urged countries to agree to restrict the level of CO_2 gas emission in the atmosphere, which scientists agreed would provide time to develop further policy options. While a significant number of countries pledged to limit their gas emissions, a few of the largest per capita gas emitters did not participate in the Protocol. Furthermore, even some of the Protocol's signatories failed to achieve the gas emission targets.

The Copenhagen summit was due to develop a comprehensive policy agenda and a legally time-bound protocol for limiting greenhouse gas emissions. Unfortunately, it did not eventuate and the world remains divided. Some small island states are increasingly threatened by rising sea levels. The reasons for failure are complex. Any action to limit greenhouse gas emission in the atmosphere requires everyone's cooperative effort. Unfortunately, Copenhagen yet again proved that, despite the hyperbole ("one world, one humanity"), we often behave selfishly.

Drainage basins and watercourses have shaped human settlement patterns for centuries. The sea played an important role, as adventurers discovered new worlds and traders explored new markets. The colonial maritime powers, namely Portugal, Spain, France, Holland, and Britain, redrew the global political map, to something similar to what we know today. The colonial powers needed to preserve a safe and secure sea route between their colonies and metropolitan bases. This contributed to developing laws and conventions governing the uses of international watercourses.

The situation is different with riparian waters. Even today no formal convention exists, although in 1997 the United Nations established the *UN Convention on the Non-navigational Uses of International Watercourses*, which includes rivers, lakes, and streams. Burundi, China, and Turkey, all upper riparian countries, opposed the convention. Of the eight South Asian countries, India and Pakistan abstained and others voted in favor of it. The International Court of Justice has used some provisions of the dormant convention in its decision on riparian water-sharing disputes.[1]

The convention is a prescriptive document and does not provide any binding mechanism for dispute resolution. It establishes guidelines and provides basic principles for resolution of water-sharing disputes between countries. The convention includes only groundwater connected to surface water, and does not extend to confined groundwater, despite confined groundwater being a major source of freshwater in many countries.[2] The issue has gained importance since the confirmation that transboundary groundwater is not an isolated entity, but instead linked with other water resources,[3] with the consequent possibility of being contaminated from both areal and point sources.[4]

Many, including the United Nations in 1989, have indicated the need for a comprehensive definition of freshwater. A major problem is that while surface water monitoring is deficient in many countries, the situation with respect to groundwater is worse. Many nations lack any proper monitoring of groundwater recharge quality.[5] The World Bank and the World Health Organization both delayed in recognizing that much of the groundwater in Bangladesh and parts of India was contaminated with arsenic.

The poor state of groundwater is an important issue to South Asia, given Nepal, Bhutan, Bangladesh, and India all have interrelated ecological, hydrological and geophysical features. A bilateral water-sharing arrangement between any two countries may affect the quantity and quality of aquifers linked to transboundary waters in the region.[6] This may encourage some countries to seek compensation or demand relevant signatories revisit any bilateral agreements reached between countries without the agreement of neighboring countries.

The South Asian Association for Regional Cooperation (SAARC) consists of eight countries: Afghanistan, Bangladesh, Bhutan, India, Maldives, Nepal, Pakistan, and Sri Lanka. Afghanistan is located between central Asia, South Asia and west Asia. In terms of politics and security, differences of policy among the various countries toward the Taliban regime have caused or aggravated estrangement and hostility between countries in the region.[7]

The issue of admitting Afghanistan within the SAARC created division among other members. In 1985, SAARC did not admit Afghanistan due to the objections of some member countries. The 2007 summit agreed to admit Afghanistan only after Nepal and Pakistan obtained China's observer status at the summit in exchange. India opposed China's observer status, as it did not want to provide its smaller neighbors with the opportunity to play the China card in SAARC matters.

Since gaining observer status, China has adopted a proactive and enthusiastic stance to extend cooperation to SAARC countries on issues such as poverty alleviation, disaster relief, and mitigation. China's determination to win over some SAARC countries has put India in an uncomfortable position.[8] One researcher argued that both India and China are trying to reconcile their identity as members of the global South and the developing world with their emerging power status, or de facto great power status in the case of China.[9]

There might be some truth in this given that, despite the rhetoric, the India-China relationship is lukewarm and the border dispute remains unresolved.[10] When China resolved her border dispute with Nepal in 1961, by conceding ten of the eleven villages that were in dispute, India denounced the action.[11] China has staked claim to India's Arunachal Pradesh in the northeast. In a complex global political world, one cannot assume that Nepal would not play China off against India within the SAARC.[12]

Since China's border conflict with India in 1962, China's support for Pakistan has remained undiminished in all matters relating to India. China's axis with Islamabad has formed a remarkably constant element in South Asia's power relations. China counts as the most important external influence in contemporary Indo-Pakistan tensions, even though it aggravates this rivalry less now than during the 1970s and 1980s.[13] Pakistan publicly eulogises its relations with China. This was clear during Pakistan's dispute with India on the construction of the Baglihar dam.[14] Pakistan further confirmed that the dispute affected its recent talks with India aimed at normalizing their relationship. Pakistan shares common values with China and the two provide mutual support to each other, "never disagreeing."[15] The media has reported that Pakistan foreign minister Shah Memood Qureshi said in Islamabad that Pakistan had given a blank check to China to negotiate with India on its behalf so as to deal with the aftermath of the Mumbai terror attacks in late November 2008.[16]

The anti-Indian sentiment is common in Pakistan; the "water scarcity" issue provides an additional opportunity to worsen the situation. It has been reported that the militant jihadi group *Jamaat-ud-Dawah* recently ignited passions in the people at a gathering in Islamabad, alleging that India was determined to starve Pakistan of water.[17] The group's head is reported to have told a gathering in Lahore that the next war between India and Pakistan could be fought over water if India does not stop what he described as "water terrorism," and he called for a jihad against India and asked the Pakistan government to prepare its citizens for this.[18]

China, a water-scarce country,[19] will exhaust its exploitable water resources by 2030.[20] China's push for development remains unabated, as local leaders stubbornly embrace the mantra of "development first."[21] In this context, China's observer status has political implications. First, in any future negotiations concerning the sharing of the Brahmaputra waters with India, China would expect to receive the support of those countries to which it is closely aligned. Second, smaller neighbors receive in exchange an opportunity to restrict India's influence on them. At the same time, China's status is enhanced as a regional and global economic and political actor.[22] As climate change threatens supplies of freshwater, nations jockey to control scarce sources.[23]

China is actively promoting itself as a global economic and political power, both overtly and discreetly. Individual circumstances determine the type of methods adopted in each case. China's influence has grown in East Asia and other parts of the world.[24] Its influence in Africa and the Middle East has increased enormously.[25] China is increasingly playing a more assertive role in global affairs, including in the United Nations.[26] It is also increasing its economic linkages with resource-rich countries.[27] While it would be premature to conclude that China's intentions are hegemonic, it is seeking to ensure its sources of raw material remain safe.

China and other outreach countries such as Brazil, India, Mexico, and South Africa, possess a range of economic, military, and political powers and have some capacity to preserve and promote international order, regionally and globally. These countries share a belief in their entitlement to a more influential role in world affairs.[28] As China moves ahead of its rivals, it may develop a hegemonistic outlook.

Of the mainland South Asian countries, Nepal, Bhutan, and Bangladesh are small; while Afghanistan is trying to establish itself as an independent, well-governed state after the trials of Taliban leadership. China's presence provides them with an opportunity to take independent positions, irrespective of India's wishes, on issues within the SAARC. People in these countries are living in abject poverty. The presence of India and China could be either a boon or a curse, depending upon how each of the countries individually positions itself in relation to these two neighbors so as to take maximum advantage of the rivalry between them. It is logical for China to grab these opportunities and use them to its advantage.

Barring Nepal and Bhutan, all mainland countries in South Asia are on the verge of becoming water-scarce countries. In this environment, security issues and allied strategic concerns, in terms of internal contradictions and perceived defense and economic needs, are different. The common issue over which none of these countries has any control is the impact of climate change on their physical environment and water availability.

India, unlike China, is both an upper and a lower riparian country. India does not have sole control of any of the major perennial water sources. Its policy response on managing the riparian waters will come under intense scrutiny by other countries. Water is a political issue that countries often use to divert public attention away from their own policy failures.

By examining these issues, this book hopes to assist countries to address riparian water-sharing issues in proper perspective, and to avoid the grim future for humanity predicted in a recent UN report.[29]

The book argues that the future remains uncertain for South Asia because, unlike Europe, it does not have a history of working cooperatively to develop regionwide policies that benefit all. The growth of fundamentalism in the form of the Taliban and al Qaeda (and its attendant complex web of international networks) makes it more difficult for nation-states to develop regionwide policies. In facing such a threat, they may ignore their long-standing mistrust of each other; yet failure to address the critical issues of climate change and water crisis is no option:

> A society no longer able to reconstruct itself is a society deprived of its capacity for autonomous reproduction. There is injury to the interests, there is insufficient capacity to undo the injury. . . . And the same applies to social space. . . . The history of civilization seems to indicate more than clearly that it is the moment, when the rulers think that they have developed the only correct idea and put it into practice with a social order with only one sociotope, that the end of that civilization is in sight.[30]

NOTES

1. Salman and Chazournes 1998: 169.
2. Salman and Uprety 2002: 27.
3. Borole et al. 1978: 181–201; Nair et al. 1978: 803–26. See also, International Atomic Energy Authority's annual report "B.1.4: Using Isotope Techniques to Study Water Resources" (Vienna: International Atomic Energy Agency, 2004).
4. Freeze 1997: 149.
5. Revenga et al. 2000: 8.
6. Blomquist and Ingram 2003: 162–69.
7. Zhao 2005: 20.
8. "China Goes into Overdrive," *Statesman Weekly*, April 7, 2007.
9. Chin 2008: 96. See also, Nafey 2008.
10. Hiro 2010.
11. Chung 2004: 104.
12. Chapman 1995b: 20.
13. Clad 1996: 183.
14. Pakistan prime minister Saukat Aziz at a meeting with editors of the *Asia News Network* in Davos, Switzerland, "Dam Disagreement Affected Talks," http://www.thestatesman.net/page.news.php?clid=1&theme=&usrsess=1&id=66973 (accessed January 29, 2005).
15. van Kemenade 2008.
16. *Statesman*, January 21, 2009.
17. *Times of India*, February 16, 2010.
18. "Next Indo-Pak War Could Be Fought over Water: Hafiz Saeed," *Times of India*, April 13, 2010.
19. United Nations Development Programme, *Human Development Report 2007–2008: Fighting Climate Change: Human Solidarity in a Divided World* (New York: Palgrave Macmillan, 2008), 97; http://hdr.undp.org/en/reports/global/hdr2007-2008/.
20. Buckley 2007.
21. Yardley 2007.
22. Sperling 1998: 5.

23. Chellaney 2007.
24. Drezner 2007: 46.
25. Oster 2006. See also, Thomas Cargill, manager, African Program, Chatham House (London), in an interview with Phil Adams on the ABC's *Late Night Live* program, November 27, 2006; http://www.abc.net.au/lnl.
26. Economy 2007; Cohen 2007: 153.
27. Tisdale 2008: 21; Klare 2010.
28. Hurrell 2006: 1–2.
29. The fourth *Global Environment Outlook*, United Nations Environment Program, Nairobi and New York, October 26, 2007; http://www.abc.net.au/news/stories/2007/10/26/2070761.htm.
30. Galtung 1985: 19, 21.

Chapter One

Climate Change and the IPCC

Just as all matter has evolved down the centuries, climate is constantly changing. We experience some of these changes, but most occur silently. Almost every aspect of our surroundings is influenced and affected by the changing climate, and these changes, in turn, also continuously induce further changes around us.

In modern-day parlance, climate change usually refers to the phenomenon known as global warming or anthropogenic global warming. Most physical changes that have contributed heavily to climate change occur continuously. These elements are tectonic plate movements, variation in solar output, volcanic eruptions, alterations in Earth's orbit and ocean temperature and last, but not least, human-induced changes.

Of all these elements, modern-day concern mostly focuses on human-induced factors. The debate around the impact of human-induced activities on Earth's climate change, however, remains a contentious issue. Globally, human knowledge concerning the creation of the universe, its unique astrophysical characteristics, and celestial mechanics is far from accurate or complete. Despite such failings, our knowledge about the universe has improved exponentially in recent decades, though it remains far from ideal.

Scientists are uncovering the mysteries of nature, and these discoveries are helping us to better understand the climate change phenomenon. For example, some Chinese scientists now consider that the problem of geological instability and frequent landslides are problematic enough to make any huge hydraulic projects difficult and even impossible.[1] The National Aeronautics and Space Administration (NASA) found that large volumes of water stored by dams have shifted the angle of the Earth's axis by approximately 60 cm since 1950.[2] No doubt such large masses of stored water also have ecological, climatic, and health consequences.

India has experienced reservoir-induced earthquakes at Koyna and Bhatsa, but only one of these had seismicity greater than magnitude of 4.0 on the Richter scale.[3] The Koyna earthquake of December 11, 1967, measured 7.5 on the Richter scale, while Bhatsa dam is on the west coast of Maharashtra (about 90 km northeast of Bombay and about 200 km north of the Koyna Dam). Most Indian dams were created in the latter half of the twentieth century, and very little is known about their affects on Tectonic plate movement, river morphology, and changed river courses, and their ecological implications. Another example is the Koshi River, which descends from Nepal into India. This river has moved about 130 km west in the last century, following completion of the upstream Koshi dam.[4]

Although climate skeptics for the most part refuse to accept that human-induced factors are responsible for climate change, scientific consensus remains that human activities have directly contributed to the rapid rise in average global temperature over several decades.

No single discipline of science can fully explain the climate change phenomenon. Understanding climate change requires a clear understanding of the research outputs of physical and earth scientists, economists, sociologists, and other specialists. Each of the climate-related disciplines contributes to a better understanding of the climate change phenomenon. Climate change believers have made it clear that their findings are based on our current understanding of all contributing factors and that, as and when additional information becomes available, they will be able to update their findings. They insist that the existing situation is too critical to ignore.

It is becoming increasingly clear that the climate skeptics have not understood this phenomenon and, in their eagerness to discredit the work of climate change believers, they have constructed their arguments on questionable assumptions and methodology. More importantly they have used arguments often followed by discredited vested interests. There is enough evidence available in support of this position.[5] Royal Society president Martin Rees commented, in a recent wide-ranging interview, that he saw no particular problem with climate change, and one just has to decide who is serious and who is misguided; it becomes a matter of evidence.[6]

In an open letter, 250 members[7] of the prestigious US National Academy of Science wrote,

> We are deeply disturbed by the recent escalation of political assaults on scientists in general and on climate scientists in particular. . . . Assaults on science and, more disturbingly, on climate scientists by climate change deniers are typically driven by special interests or dogma, not by an honest effort to provide an alternative theory that credibly satisfies the evidence. We also call for an end to McCarthy-like threats of criminal prosecution against our colleagues based on innuendo and guilt by association, the harassment of scientists by politicians

seeking distractions to avoid taking action, and the outright lies being spread about them. The good news is that smart and effective actions are possible. But delay must not be an option.[8]

BIRTH OF THE IPCC

The Intergovernmental Panel on Climate Change (IPCC) was established by the World Meteorological Organization (WMO) and the United Nations Environment Programme (UNEP) in 1988 as a lead agency to investigate climate change issues. To date it has issued four reports, the latest in 2007; the next report is due in 2014. Each report has asserted with increasing certainty that the world is warming. It has assessed with 90 percent certainty that human-driven greenhouse gas omissions are largely to blame. As the IPCC's charter requires it to remain policy neutral, it does not make any policy recommendation, nor does it initiate any.

Generally speaking, the world accepts that global warming is the most serious of all environmental threats that we face in the new millennium. In its Fourth Assessment Report the IPCC issued a blunt warning. It claimed that temperatures are the hottest in a thousand years.[9] NASA's Goddard Institute of Space Studies and National Climatic Data Center concluded that 2005 was the warmest year since the late nineteenth century, surpassing the previous highest in 1998. The WMO and the Climate Research Unit estimated that the 2005 temperature was marginally behind that of 1998. The IPCC indicated that without swift remedial actions, the world was on course for a 1.8°C–4.0°C increase in warming, and by century's end, hundreds of millions could be at threat from drought, flood, storms, rising seas, disease, malnutrition, and homelessness.

The rate of warming has been about 0.2°C per decade on average in the past thirty years, but the warmest ten years on record has been since 1990. Other studies confirm similar trends; one study found that nineteen of the hottest twenty years on record have occurred since 1980.[10] Some weather experts now predict that 2010 may turn out to be the hottest year on record, citing previous experience that a strong El Niño climate pattern contributed to the record temperature in 1998.[11] Recently, a team of international scientists has updated the scientific data on global warming and claimed that the findings indicate a far worse situation than predicted by the IPCC. The data clearly refutes the position adopted by climate skeptics.[12]

CLIMATE CHANGE IMPACTS

Climate change touches all aspects of human life and its impacts are enormous. This enormity is real, as environmental aspects are intrinsically interrelated and affect all ecosystems, adversely impacting on humans and their activities.[13] Extreme weather events will impose significant costs, with some estimating up to US$100 billion annually,[14] but this amount could be higher if remedial measures are not adopted to address the basic cause of climate change—the greenhouse gas emission by human activities.

The impacts of climate change are daunting for all countries, but those worst affected are poorer nations that mostly lack resources to overcome the consequences of transformation. For example, between 1900 and 2005, an estimated two billion people were affected by drought and an estimated more than ten million people died. Ninety-seven percent of the deaths occurred in eight droughts, and mostly occurred in India and China.[15]

Some skeptics argue that global warming may not be as critical as scientists make it out to be.[16] Differences aside, scientists have sufficient information to make most predictions with a high degree of certainty. For example, researchers recently found that anthropogenic in atmospheric CO_2 is driving fundamental and unprecedented changes in the chemistry of the oceans. This has led to changes in the physiology of a wide variety of marine organisms, and consequently the ecology of the oceans. The authors argued that ocean conditions are already more extreme than those experienced by marine organisms and ecosystems for millions of years, and emphasized the need to adopt policies that drastically reduce CO_2 emissions.[17]

There is less hard data on global warming issues, as the science of climatology is far from perfect. However, most scientists do argue that it is essential to spread the search for climate facts beyond the narrow realms of peer-reviewed journals. India's Shiva argues that more emphasis should be given to "people science."[18] For example, often farmers have a better understanding and knowledge than scientists about many natural phenomena, and which narrowly focused research may fail to unearth. IPCC's failure is that it did not make it clear that, in the absence of hard data, some forecasts were based on government and advocacy group reports. The lead author of the first IPCC report came close to the truth that "a chronic problem is that on the whole area of impacts, getting into the realm of social science, it is a softer science. The facts are not as good."[19]

SKEPTICS AND THE IPCC

Since the publication of the leaked e-mail (from the data base of East Anglia University's Environment Unit), climate skeptics have virtually declared war against the IPCC. The IPCC and scientists associated with the organization agree that mistakes and oversights are unacceptable, but climate skeptics have found it convenient to selectively use the leaked material to strengthen their argument. A number of issues have ignited the current debate: first, accuracy of the IPCC's assertion predicting that the Himalayan glaciers could disappear by about 2035; second, the impact of the diminishing Amazon rain forest; third, the link between global warming and natural disasters (hurricanes and floods); and fourth, confusion about how much of the Netherlands is below sea level.

The unfortunate outcome of the error in the date of melting is that it detracted peoples' attention from the fact that the Himalayan glaciers are melting rapidly.[20] The IPCC admitted that the claim was based on an unpublished report that had not been subjected to routine scientific scrutiny.[21] The IPCC chair, Rajendra K. Pachauri, reiterated the overall conclusion that glaciers gather water from seasonal snow packs, but the process with the Himalayan glaciers is different. The Indian Space Research Organisation (ISRO) found that, overall, there has been 16 percent de-glaciation in the Himalayan basin.[22] Given the rising temperature, we cannot ignore the possibility of a faster rate of decline.

The World Glacier Monitoring Service analyses data for just over ninety glaciers and notes the continual decrease in their average mass balance. Professor Ian Allison, a veteran glaciologist has argued that "there are a very small number of that are monitored pointing to 100 globally for which there are regular 'mass balance' measurements that reflect how much a glacier grows or shrinks from one year to the next."[23]

The IPCC should have warned readers that the information available from observations of glaciers from different parts of the world may not be comparable to the glaciers of the Himalayas. The way they gather and lose water is totally different from European Alpine glaciers, given the differences in geological, meteorological, and physical characteristics between the two regions. Glaciologist Allison confirmed that glaciers in one region can all apparently behave differently in response to the same climate signal, because the fluctuations that occur in the front depend on how long it takes to transfer the mass from the top of the glacier to the bottom.

The IPCC pointed out that it did refer to two published articles, one being a dissertation written by a professional mountain guide and climate change campaigner while studying at the Berne University. It certainly raises a question of when a scientist is not a scientist. It is possible that the dissertation was less vigorously researched, but a dissertation on the subject certainly could not be entirely ignored by scientists in such complex matters.

The IPCC also agreed that two reviewers who checked drafts of the IPCC report urged greater caution in proposing a link between climate change and disaster impacts, which was ignored. The IPCC says that it is reassessing the evidence and will publish a report after this process is complete.[24]

Another bone of contention is the future of the Amazon forests, which skeptics have dubbed as "Amazongate." Simon Lewis argues that the IPCC statement is basically correct but poorly written and bizarrely referenced. He argues,

> The most extreme die-back model predicted that a new type of drought should begin to impact the Amazon, and in 2005 it happened for the first time: a drought associated with the Atlantic, not Pacific sea surface temperatures. The effect on the forest was massive tree mortality, and the remaining Amazon forests changed from absorbing nearly two billion tonnes of CO2 from the atmosphere a year, to being a massive source of over three billion tonnes.[25]

On the Netherlands issue, the IPCC claimed that the figures were supplied by a Dutch government agency. It, however, appears that the statement reflects an imprecise expression and does not alter the report's conclusion. It said that the figure had been used to mean "either the area below the highest sea-level reached during storms, or the total area of the country that is prone to flooding from the sea and rivers. Therefore, a preliminary analysis suggests that the sentence discussed should end with: because 55 percent of the Netherlands is at risk of flooding."[26]

Scientists have not denied identified errors, but any errors are minor and, barring the Himalayan glacier conclusion, the thrust of the report remains unchallenged. Confusion concerning these glaciers may be incorrect, but the fact remains that they are melting more rapidly than glaciers in other parts of the world. It is worth noting that the Himalayan glaciers exist in the monsoonal climatic region and gain mass from precipitation during the same warm session in which they lose mass from melting, making their behavior complex.[27]

The issue is the quality of the assessment of the available information on climate change phenomena and the capacity of the scientific community at large to establish the causal relationship between events and their findings, rather than the issue of unanimity per se. It is on the basis of these collective interpretations that governments and the global community decide what action to take, and how to minimize any negative consequences. An article in the *Economist* noted,

> Some scientists think that the planet is already on an irreversible journey to dangerous warming. A few . . . sceptics the problem will right itself. Either may be correct. Predictions about a mechanism as complex as the climate cannot be made with any certainty. . . . As an insurance policy against a catastrophe that may never happen, the world needs to adjust its behaviour to try to avert that threat . . . [and] broadly agree that greenhouse-gas emissions can be curbed without flattening the world economy.[28]

The global scientific community has publicly supported the direction and most of the IPCC predictions, and has also highlighted the areas where the agency needs to strengthen its work procedures and practices. Even US government officials have argued that the bulk of the IPCC reports have merit. Britain's former chief scientific adviser Professor David King and chief of the US National Oceanic and Atmospheric Administration has argued that "the vast majority of conclusions in the IPCC are credible, have been through a very rigorous process and are absolutely state of the science, state of the art about what we know of the climate system."[29]

In blunting the skepticism of the climate skeptics, highly respected scientists argue that hard sciences support most of the IPCC conclusions. For example, it is well established that there exists a very close link between sea temperature and atmospheric climate change. There is no disagreement that the Arctic Ocean ice is melting and the glaciers receding worldwide, and that the subtropics are expanding, resulting in more extreme variations in climate globally.[30] As mentioned earlier, scientists are constantly updating their information base. Consequently, the foundations of their earlier assertions are strengthening in most cases. For example, scientists recently have discovered a deep ocean current with a volume equivalent to forty Amazon Rivers in the ocean sector of the Southern Ocean. This is an important step in understanding the global network of ocean currents that influence climate now and in the future.[31] This is a big leap in understanding the relationship between the ocean temperature and climate change.

Greenland is now losing ice rapidly, and Antarctica is beginning to lose ice at a rate of about 150 cubic kilometers a year. Eventually the situation will reach a point where the dynamics of the system begins to take over. When the ice sheets begin to collapse, the tipping-point will have passed, causing ice sheets to end up in the ocean. Further, ocean warming can mean methane hydrates begin to release methane (a strong greenhouse gas), which increases global warming and causes sea level to rise.

Scientists have found that, in stable climate times, the amount of heat coming into Earth's system is equal to the amount leaving it, but it is not happening and half of the gap remains unaccounted for. They have concluded that humans have contributed to this imbalance by pumping climate-warming greenhouse gases into the atmosphere. This is evident from warming ocean waters, melting polar ice and the like. They have further concluded that this heat will come back to haunt the world sooner or later.[32]

Michael Oppenheimer, one of the earliest climate scientists argues that there certainly exists a degree of uncertainty surrounding climate change issues, and that no one should assume that climate scientists are providing absolute answers. He believes that scientists working on the IPCC program consider their job to be making both the public and policy makers aware (as transparently as possible) of this climate change uncertainty, and what they do or do not know.[33] It is unlikely that there will be unanimity between most scientists and those who remain skeptical about the climate change issues. Oppenheimer further states,

> In order to ensure against subconscious or unconscious bias, IPCC has a multi-layered review process where the documents in draft form go out to a small group of scientists to look and scrub every fact. Then they go out to a much bigger group of scientists, hundreds of them to scrub down every chapter. Then they go to the governments, who in turn them over to their own scientists to scrub it down. So that by the time you get to producing a final document, it's not only been reviewed three times, but there's what's called a review editor who sorts through all comments.[34]

James Hansen, a well-known NASA scientist, claims that during the ten thousand years between the last ice age to the present interglacial period, with the disintegration of the North American big ice sheet, sea levels rose 5 m every century; and 1 m every twenty years for several centuries. Once an ice sheet begins to melt and disintegrate, the situation deteriorates rapidly.[35]

He highlights that climate change progress does not follow a linear trend, and this fact makes climate change phenomenon an extremely complex issue. In support, he points out that it took the ocean a long time to warm up (being 4 kilometers deep), and that it takes ice sheets a long time to start moving. Ice sheets are also very thick and have significant inertia. He further argues that the data used to determine the temperature change over the last century or so was available for a long time, and during this period the skeptics neither expressed concern nor raised questions; and that now they try to find small things and question the integrity of scientists.[36] Their reactions are not backed by research findings, but they have accused the IPCC of gross irregularity.[37] None is arguing that the skeptics have no right to raise issues with the IPCC, but, rather, that hardly any research-based argument has been forthcoming in support of their positions.

A large number of scientists, environmentalists, and activists have also come out strongly in support of the IPCC. A key author of the IPCC's 2001 and 2007 reports argued that questions have been raised about only two paragraphs in a 1600-page document, and that people ought to be talking about the other 1599 pages with which nobody had found any problem.[38] Certainly there were omissions, and even some internal structural deficiencies within the IPCC that could have led to this situation, but we need to remember that the review process in this instance included forty thousand comments.

While it is undeniable that the few omissions and oversights have raised doubts in the minds of at least some, the fact nonetheless remains that most scientists agree that the thrust of the IPCC findings remain sound. It is understandable that skeptics on any issue will wait for opportunities to torpedo the issues they do not like. This is particularly the case for issues with a strong vested interest (in this case the fossil energy lobby), where they will spare no effort to discredit any report that goes against their interests.

The IPCC, various national academies of sciences and the UN administration unequivocally and collectively agree that trust and transparency are basic to all scientific findings. These two aspects need to be preserved and promoted at any cost. The president of the US National Academy, and the president of the British Royal Society were forthright in their argument that scientists needed to be more open about their findings, and that people expect them to do things more in the public light and that scientists have to get used to this prospect. United Nation's secretary general Ban Ki-moon acknowledged the scientists' position and requested the Inter Academy Council in Holland to undertake an independent review of the IPCC's procedures and processes.[39]

In response to the criticism by climate skeptics, it is argued that even if a fraction of the scientists' predictions were to come true, the collective impact of climate change globally would be substantial in terms of increased poverty, hunger, and human displacement, particularly in South Asian countries. Singularly or collectively (or in combination), these could trigger a variety of conflicts and enormous suffering, both within and between countries.

The importance of understanding the nuances of greenhouse gas emission and climate change phenomenon is beyond human comprehension. Monetary and financial considerations alone are insufficient to address climate change issues.[40] Rising sea level caused by climate change would force hundreds of millions of people to move from their usual habitat. Such forced movements create global conflicts that endure for a long time, with little available to stop those movements[41] A recent study on national security and climate change in the United States argued, "We will pay for this [climate change] one way or another. We will pay to reduce greenhouse-gas emissions today, and we will have to take an economic hit of some kind. Or we will pay the price later in military terms. And that will involve human lives. There will be a human toll. There is no way out of this that does not have real costs attached to it."[42]

There are similar examples in Asia's history. Climate historians found that droughts provoked two periods of conflict in Southeast Asia in the fifteenth and eighteenth centuries.[43] In a modern-day context, climate disturbances will create political and military tensions, and encourage a blame game by unscrupulous political leaders who, even in normal circumstances, do not often hesitate to use every trick to further their own interests.

Skeptics have not been able to substantiate their criticisms on climate change issue in any meaningful way. Though the uncertainties in climate change cannot be denied or ignored, neither can the fundamental argument underpinning the climate change debate. As argued in an *Economist* article: "In any complex scientific picture of the world there will be gaps, misperceptions and mistakes. Whether your impression is dominated by the whole or holes will depend on your attitude to the project at hand. . . . The doubters are right that uncertainties are rife in climate science. They are wrong when they present that as a reason for inaction."[44]

HOW THE IPCC OPERATES

Each IPCC working group is headed by two scientists, one each from a developed and developing nation. Each working group is supported by a large number of scientists, known as lead authors, who, in turn, are supported by reviewers.[45] Ove Hoegh-Guldberg has significant experience of, and association with, the IPCC. He confirms that, although minimal mistakes or oversights may exist, the methodology followed and working mechanism adopted by the organization were the best in the circumstances.[46]

There are various stages of review. The first round involves the working groups picking over the text: hundreds of eyes and qualified expert opinions. Often it takes months or years and involves pedantic haggling over detail, but it always uses peer-reviewed literature as the base. The documents are sent to signatory governments for review on completion of this process.

The reports are then placed on the Web for open comment. The lead and contributing authors are obliged to respond to each comment, irrespective of its merit or value. The comments are then independently reviewed (by referring to peer-reviewed literature) to ensure all comments have received an objective response. Hoegh-Guldberg confirmed that he personally had to respond to eighty-seven comments on a relatively small contribution to the Australian and New Zealand working group 2 of the IPCC report in 2007. He found most of the comments to be conservative. All public comments on draft reports received written replies. In comparison, just one of the 2007 working groups elicited more than thirty thousand comments from its open public review process.

The IPCC processes provide opportunities for governments, research institutes and even individuals worldwide to directly point out any deficiencies or shortcomings in the draft report. The Center for Environment and Geographic

Information Services in Bangladesh has recently claimed that "even if sea levels rise a maximum one metre in line with the IPCC's 2007 predictions, most of Bangladesh's coastline will remain intact."[47]

The aforesaid Bangladesh study raises a very important issue of why scientists in Bangladesh and India did not raise this issue immediately following publication of the IPCC report. Further, even before the IPCC report, two other reports were published (one by former US vice president Al Gore, and the other by the Lowy Institute in Sydney, Australia). Both reports argued that thirty million to sixty million people will lose their habitat in Bangladesh and India because of the impacts from climate change. It is odd that this issue should be raised after a gap of so many years. The situation in the Indo-Gangetic Plain and other deltas does not provide much scope for confidence. Further, Atiq Rahman, the IPCC's representative in Bangladesh, also did not raise this issue, perhaps indicating that either he was not confident that the report could conclude otherwise or that it was irrelevant.

Andy Pittman, a lead author on the 2001 and 2007 reports, acknowledged that there was much that scientists did not know about how the earth and climate work. Scientists may be mistaken on some things, but they are eager to test and retest emerging data. If anything, they are by nature a conservative bunch, and "all good scientists are sceptics to the core." Pittman's response to the skeptics confirms the strong ethical commitment of the scientists working within the IPCC system. He argued that the skeptics have neither researched nor published any climate science related paper in peer-reviewed journals: peer review is a significant part of the scientific process. He emphasized that climate science "is not about opinions. It is about what is provable and disprovable. There are no scientific publications that provide serious challenges to the IPCC conclusions. Not one. But there are literally thousands that support the (IPCC) conclusions." Martin Manning, who headed the technical support group for the 2007 IPCC report, made similar comments.

CLIMATE CHANGE: ECONOMICS AND SOCIAL IMPACT

As mentioned earlier, the impacts of climate change are widespread and varied. It is practically impossible to identify every event that is impacted by climate change. For example, changes in the frequency and intensity of heavy precipitation, economic and physical consequences of glacial retreats, rising sea level, impacts on vegetation, food supply, and health issues are all significant issues and difficult to measure accurately at this stage.

Generally speaking, the business community always adopts a hard-nosed approach on noneconomic issues, but they are usually very pragmatic when dealing with such issues. At the 2009 World Economic Forum, the business community agreed that, even with the economic crisis capturing the attention, there may be a slowdown on regulations concerning the environment. There was consensus that both developed and developing countries will have to make efforts to pursue environmentally friendly policies; and that developing countries may be held to a less stringent standard than developed countries.[48]

It became clear when the head of Australia's peak coal body, usually a key constituent of the fossil energy lobby, stated in an interactive discussion group that any acceleration in global warming would demand drastic action.[49] His public pronouncement confirms that business often understands complex issues and takes a long-term view of their own economic interests.

As indicated earlier, climate change and blanket mitigation prescription would severely restrict the capacity of developing and less developing countries' capacity to achieve rapid economic growth, because to achieve high economic growth rate major structural changes would be required to the operation of their economies. This is the reason why these nations have demanded long-term funding and technological assistance to adjust their economies in the new economic environment. Yet, developed economies will find it much easier to adapt to a green economy should they adopt appropriate measures to cash in on the new markets generated by a green economy. The new policies certainly need to be comprehensive enough to cover energy, agriculture, and other related sectors.

CLIMATE CHANGE AND PUBLIC CONCERN

Worldwide, the public was indignant that the Copenhagen Summit had failed to achieve a more positive outcome. The public was jolted with the revelations alleging that the IPCC cheated, although this was not said in so many words. People were confused and wanted to be made aware of the real reason for the failure. David Freestone argued that it was a difficult concept for the people to grasp, as they were receiving mixed messages.[50]

Climate change certainly is no longer a technical issue; the skeptics have successfully hijacked the agenda following the widespread publicity of the lapses found in the IPCC. This is a tragedy, because the debate has focused on the lapses rather than the main issues. The political position adopted by two influential bodies (namely, the US Senate and the Australian Senate) has certainly helped to create confusion in the minds of a large sector of the public. By and large, the European Union has been at the forefront of the climate change issue, and has certainly adopted far-reaching measures in their environment policies when compared to many other countries.

The major sticking point remains the disagreement between the United States on one side, and China and India on the other. The most contentious issue is how to measure the amount of allowable gas omission: per capita or total. It is almost certain that none of the main protagonists will back down on this issue, as all of them feel that such a move means disadvantaging their respective economies. The economic downturn, phenomenal and sustained growth in the Chinese economy, and short election cycles in countries such as the United States and Australia prevent political leadership from adopting policies that might adversely impact on their economies in the short term.

The public's confusion is widespread. For example, a number of callers in radio talk-back shows in countries such as Australia publicly sought guidance on how to get facts about whether there is a human-induced global warming. In summary, callers argued that they have not read or did not wish to receive the IPCC reports, but desired independent information. It certainly conveys the message that at least some did not consider the IPCC reports independent and that they may contain unreliable information.

Even in developing countries such as India and Pakistan, a 2007 World Economic Forum survey found that in India only 16 percent identified environment as one of the top five issues out of thirteen issues. In Pakistan it was not even one of the top five issues. In both these countries global warming and water crisis are already major issues, given their economic survival is significantly dependent on the success of their respective agricultural sectors.

The word independent implies that the concerned issue is free of any bias. Climate change issues cannot be assessed without the informed contribution of a large number of scientists and professionals from various disciplines and professions. The IPCC achieves this objective by gathering hundreds of scientists from all over the world.[51] To ensure that their assessments are objective, and not biased, scientists are not allowed to offer recommendations with their assessments.

Globally there have been cases where scientists (particularly in the medical field) have been accused of falsifying or manipulating data, the consequences of which have proved dire.[52] Accusations have taken various forms. The Federal Opposition in Australia has dubbed the IPCC reports as alarmist and extreme political ideologues that are helping to de-industrialize modern society.[53] Such a belief injects a germ of mistrust between developed and developing nations, implying that the nonindustrialized countries are conspiring to damage the industrialized countries by destroying their economic bases. This has complicated the debate, as perceptions are often more powerful than logic, particularly in an emotionally charged complex issue such as global warming and how to mitigate it.

It is unfortunate that the alternate government in the country used such language in opposing the government policy, particularly when the two principal scientific agencies in the country have been arguing that climate change has been real for some time. The Commonwealth Scientific and Industrial Research Organization (CSIRO), Australia's principal scientific agency, and its Bureau of Meteorology have been researching the climate change issue intensively for some time. Principals of these two organizations have publicly confirmed that it is extremely unlikely that global warming could be explained by natural causes alone. They state that evidence of human influence has been detected in ocean warming, sea level rise, continental-average temperatures, temperature extremes, and wind patterns.[54]

It is unfortunate that, at the height of the climate change debate, the world has experienced an unprecedented financial crisis leading to near collapse of the global financial order and the banking system, which has led to soaring unemployment. This has provided a superficial opportunity for many to fail in their understanding of the real cause or causes of the crises. Lord Monckton, a staunch climate skeptic and an advisor to former British prime minister Margaret Thatcher, went further by directly arguing that any attempt to significantly address climate change would cost trillions of dollars, which Western governments and their citizens cannot afford, and would bankrupt their economies. He did not offer any real argument for his opposition to the IPCC's assessment. One journalist in India has even coined terms such as "climate theology" and "shaky testament" while discussing climate change issues and the outcome of the Copenhagen Summit.[55]

The debate also provided an opportunity to Al Qaeda's Osama Bin Laden, who allegedly issued a statement arguing that the United States, by not accepting the Kyoto Protocol and by emitting more harmful gases in the atmosphere, was doing great harm to the economies of the poorer countries.

ROLE OF NGOS

While the public at large remains confused, climate activists are usually led by nongovernmental organizations (NGO). The unrelenting campaign and the commitment of NGOs toward environmental issues have hugely contributed to an increased public awareness and understanding of global warming issues. NGOs however need to be careful that, in their exuberance, they do not harm their cause. For example, environmental reporter Seth Borenstein has highlighted the pitfalls of such practices in a report from Copenhagen.[56] He refers to a group that successfully argued to a gathered audience (that included scientists and church leaders) that a figure represented a line between habitability on earth and an extremely desolate future. The figure refers to 350 parts per million of carbon dioxide in the atmosphere, a figure they claim is the highest concentration the world can manage without sparking dangerous climate effects.

Scientists point out that reducing carbon dioxide levels to 350 parts per million would mean reversing the trend of a couple of centuries, given carbon dioxide can stay in the atmosphere for as long as a century. This example highlights the point that emotionally charged claims do more harm than good to the cause, and that NGOs need to be consciously aware of this.

The contribution of NGOs and their important role in raising community awareness on the climate change issue are undeniable.[57] Despite their activities often having a single focus, the momentum generated at the Copenhagen Summit would not have occurred without their untiring work. In the end, they believed the Copenhagen outcome to be a coup d'état against the UN system, framed by an elite; a betrayal of the poor; and an insult to expert opinion. Their collective frustration found expression in sentiments such as the following: "This decade must be seen as the decade when people woke up to the challenge of climate change and began creating a climate movement. . . . Over the next few years, as awareness grows and the effects of climate change unfortunately become more apparent, politicians will have no other choice but to fall in line."[58]

Having acknowledged the role played by NGOs in mobilizing public opinion, NGOs' tasks have, however, been made difficult by the IPCC's lapses and their initial response, although no lapses provide any reason to doubt that global warming is real, is human-made, and will create problems for the world.[59]

Scientists, climate activists, and social commentators tend to think that efforts to raise the consciousness of people on this issue have to be different. One argument gaining currency is that material consumption in some parts of the global society has reached absurd levels, it having been engineered by marketers and advertisers and assisted by the fossil fuel lobby and allied commercial interests. Behavioral issues are therefore likely to be more important than any description of exactly what is happening or might happen to the climate. There is therefore a need to generate informed debate, and to not trivialize the climate question with marginal and peripheral issues.[60]

WHAT IS THE FUTURE?

There are areas where uniform global action is certainly required, but at the same time individual countries also have to take steps to meet their specific needs. In other words, political aspects of policy issues are unavoidable, as impacts on individuals and nations vary depending on agreed policies. An article in the *Economist* noted,

> It is all about politics. Climate change is the hardest political problem the world has ever had to deal with it. . . . At issue is the difficulty of allocating the cost of collective action and trusting other parties to bear their share of the burden. At a city, state and national level, institutions that can resolve such problems have been built up over the centuries. But climate change is a worldwide worry for only a couple of decades. Mankind has no framework for it. The UN is a useful talking shop, but it does not get much done. . . . The closest parallel is the world trading system . . . but the WTO is child's play compared with climate change.[61]

People who have followed the Copenhagen debate are aware that the US role was scrutinized carefully. There are reasons for this. First, although per capita greenhouse gas emission is highest in the United States, the Summit did not endorse the Kyoto protocol. The Pew Research Center in the United States found that the issue of global warming was not even within the top twenty issues in 2009, although 57 percent thought there was solid evidence of global warming (14 percent less than the previous year).[62] Some are even inclined to believe that "climate change is a temporary aberration" and that, once carbon dioxide emission is curtailed, everything will return to normal. Climate researchers have found that the change due to any increase in carbon dioxide concentration is largely irreversible for one thousand years after the cessation of emissions.[63]

Only after the presidential election in 2008 did the United States admit that climate change was a major policy concern. It did not oppose the Bali Declaration that a revised Protocol will become effective in 2012, the main agenda of the Copenhagen Summit. More importantly, whatever the US delegation was to say at the Summit, it had to be mindful of what Congress, and especially the Senate, was thinking. Notwithstanding, observers interpreted the delegation's body language as spreading a message of "we will do better."

This became clear when one of the senior US negotiators was alleged to have said that he believed the "Climategate" issue would end up being a blip in the global warming debate. There has been a strong push back to essentially saying that we are talking about thousands of scientists around the world, peer reviewed papers (literally in the many thousands), and that we are talking about full reports. We do not consider some disputes within the scientific community to be enough, in any way, to undermine the main tenets of the science of climate change.[64] Observers consider this to be extraordinary and an important marker indicating what the scientists think about the skeptics.

With the leaking of the confidential e-mail noted earlier, the director of East Anglia University's environment unit stood by his data.[65] He stood down voluntarily, pending the outcome of the independent reviews. He said that the leaked message picked up on the word "trick" in one e-mail from 1999, and talk of "biding the decline." He further said that the first thing to point out is that this refers to one diagram, not a scientific paper. The word "trick" was merely used colloquially as in "a clever thing to do." It did not have any other meaning.[66] He released a copy of the actual e-mail which reads, "I have just completed Mike's Nature trick of adding in the real temps to each series of the last 20 years (i.e., from 1981 onward) and from 1961 for Keith's to hide the decline."

A representative of the environment unit further commented that the issue raised in one of the leaked e-mails referred to an item published in 1998 which figured prominently in the 2001 IPCC report. An article in the *Economist* made an important point that there was nothing in the leaked e-mails so far to suggest that they (researchers in the East Anglia University's Environment Unit) did not believe in human-made global warming and were making the whole thing up, as claimed by some skeptics.[67]

Some scientists argued that responses to criticisms could have been better managed by adopting a less hostile attitude and friendlier tone. By not doing so, the IPCC provided an opportunity for the fossil fuel lobby to sharpen their criticism. One can argue that Professor John Beddington, the chief British science advisor, and Professor Penny Sackett, the chief Australian scientist and others of similar view, although offering only a few comments, were guilty of inserting a political overtone to their comments.[68] By and large, most comments have similar tones as that noted in the *Economist* article, which warned against orthodoxy in this debate, given the stake in the global warming debate could scarcely be higher:

> Some scientists think that the planet is already on an irreversible journey to dangerous warming. A few . . . sceptics [believe] the problem will right itself. Either may be correct. Predictions about a mechanism as complex as the climate cannot be made with any certainty . . . [just] as an insurance policy against a catastrophe that may never happen, the world needs to adjust its behaviour to try to avert that threat . . . [and] broadly agree that greenhouse-gas emissions can be curbed without flattening the world economy.[69]

Reactions varied, depending upon each country's view of the position of their economic and physical environment. Most industrialized nations believe that any remedies will involve a huge cost to them, particularly in terms of employment and economic growth, while developing nations argue that it will significantly slow down their economic progress. Developing nations

Table 1.1. Greenhouse gas emissions: Per capita and national total in 2007

	Australia	*US*	*Japan*	*EU**	*China**	*Brazil**	*India**
Per capita emission (tons of CO_2 equivalent)	26.00	24.0	11.0	10.0	5.5	5.5	1.9
Total emission (tons billion)	0.54	7.1	1.4	5.0	7.2	1.0	1.9

Source: Reproduced from *Economist*, http://www.economist.com/specialreports/displaystory.cfm?story_id=14994872 (accessed January 23, 2010).
*Denotes 2005 data.

also argue that as the current climate crises is created, in the first instance, by the past activities of industrialized nations, it is these countries which must accept the responsibility of paying for the mitigation process.[70]

Current per capita emissions of CO_2 gas are many times higher in industrialized countries than in newly developing countries. However, because of their population size, in aggregate few newly developing countries (except for China and India) are considered to be large polluters. China has argued forcefully that these countries believe that as developed nations (the industrialized nations) emitted CO_2 gases without restriction for many decades, they cannot force developing countries to reduce CO_2 gas emissions. Such a move would also severely restrict the capacity of newly developing countries to achieve economic growth. Table 1.1 shows per capita and national level of emission in selected countries.

The current global economic and physical environment reflects historical dynamics. At the start of the eighteenth century, approximately 20 percent of the world population lived in Europe (including Russia). The European population quadrupled following the Industrial Revolution. By 1913, Europe had more people than China, and the proportion of the world's population living in Europe and the former European colonies of the United States had risen to over 33 percent.

The United Nations now projects that global population growth will nearly halt by 2050, but that there will be other dramatic shifts between 2010 and 2050. During this period, the global economic growth will be driven, among other factors, by countries with a younger demographic base and rapid urbanization in what is now known as "second world" countries. In that context, it is hardly surprising that these countries remain steadfast in not agreeing to reduce their gas emissions until they have come closer to achieving economic growth and a higher standard of living for their population.[71] Global business leaders also told the recently concluded Davos forum that the long-term interests of first and second world countries were now intertwined:

> With China and India spurring the global economy . . . [it is] predicted that the difference [in] growth rates between the developing and developed worlds "are increasingly going to become larger." The result . . . is that richer countries will "more aggressively" invest in emerging markets in order to maintain their own growth, which will be "good for the emerging world."[72]

IS THERE ANY DIRECT BENEFIT FROM GLOBAL WARMING?

Global warming may bring some benefits to certain regions, such as Canada and Russia, although small island states are facing annihilation. Scientists have found that global warming is helping trees to grow faster.[73] They are not convinced whether the benefits outweigh the costs. Russian scientists confirm some disturbing findings. They have found that, in the past three years or so, a parasite (ticks) has appeared in de-iced areas which are gradually spreading to other regions. "Tick" is generally identified as any of various bloodsucking acarids of the families Argasidae and Ixodidae, which attach themselves to the skin of dogs, cattle, and other mammals, and may transmit disease to humans. An increasing number of people in these areas have been affected by infectious diseases that were not previously in evidence.[74] While health professionals are concentrating on health issues, others remain apprehensive of long-term ecological consequences and their impact on human, other animals and plant lives.

NOTES

1. A. Chen and C. Chen, "Letter to the Editor," *Nature* 429, no. 6991 (2004): 501.
2. *Harper's Index* 4 (1997): 13 and 87.
3. Gupta 1992: 1–5.
4. Chapman 1992: 28.
5. Lord Martin Rees (president, Royal Society of Britain), "Challenges for the Next 50 Years," delivered at the Australian Academy of Science on March 25, 2010. Dr. Riley Dunlop (Oklahoma State University), William Freudenburg (professor of environment and society, University of California, Santa Barbara), Naomi Oreskes (historian of science, University of California, San Diego), and Professor Stephen Schneider (Stanford University) spoke at a symposium at the American Association of Science meeting in March 2010 in San Diego. The symposium looked at the sources and strategies to climate change science. These scientists addressed how the climate skeptics promote climate change skepticism throughout society. Transcripts

of their speeches were retrieved from http://www.abc.net.au/rn/scienceshow/stories/2010/2859986.htm (accessed April 12, 2010).

6. Rees 2010.

7. Signatory affiliations are available as supporting material at http://www.sciencemag.org/cgi/content/full/328/5979/689/DC1.

8. *Science* 2010.

9. Musil 2009: 16.

10. G. Lean, "World Is at Its Hottest since Prehistory," *Independent*, December 20, 2005; cited in Smith and Shearman 2006: 3. Also, Dr. J. Hansen, a well-known NASA scientist argued that as the world is moving into the El Nino phase, 2010 will probably be the warmest on record; http://www.abc.net.au/lateline/content/2008/s2764523.htm (December 21, 2009).

11. "No Hiding Place?" *Economist*, January 9, 2010.

12. Professor David Koroly, Melbourne University, http://www.abc.net.au/pm/content/2010/s2838057.htm? (accessed March 5, 2010).

13. Smith and Shearman 2006: 24–25.

14. Saunders and Turekian 2007; Zedillio 2007; Stern 2006.

15. Because drought impacts are so pervasive and the historical record before 1964 is incomplete, the actual number of people affected is likely to be significantly higher. See Cooley 2006.

16. Watagawa 1995: 31.

17. Pelezero, Calvo, and Hoeg-Guldberg 2010.

18. Cited in "Harrabin's Notes: Reforming the IPCC Climate Body," http://news.bbc.co.uk/2/hi/science/nature/8496365.stm (accessed February 10, 2010).

19. "Scientists Seek Better Way to Do Climate Report," *Times of India*, February 11, 2010.

20. Stephen Schneider (Stanford University), Patrick Michaels (Cato Institute), and Judith Curry (Georgia Institute of Technology) also virtually endorsed the thrust of Schneider's view. "Criticism of Climate Change Science Heats Up," *NewsHour*, March 10, 2010, http://www.pbs.org/newshour/bb/weather/jan-june10/climate_03-10.html.

21. "Pachauri Admits Mistake in IPCC Report, Rules Out Resignation," *Times of India*, January 23, 2010.

22. "Himalayan Glaciers Shrank 16 Percent in 50 Yrs: ISR," *Times of India*, March 29, 2010.

23. "Glacier Melting a Key Clue to Tracking Climate Change," *Times of India*, March 5, 2010.

24. Leake 2010.

25. Dr. Simon Lewis of Leeds University coauthored a paper on the Amazon in the journal *Science* and wrote that the Amazon forest is surprisingly sensitive to drought. "Harrabin's Notes: IPCC under Scrutiny," January 30, 2010, http://news.bbc.co.uk/2/hi/science/nature/8488395.stm.

26. "UN Climate Panel Admits Dutch Sea Level Flaw," *Times of India*, February 16, 2010.

27. "Glaciers and the IPCC: Off-Base Camp," *Economist*, January 23, 2010.

28. "Getting Warmer," *Economist*, December 3, 2009, http://www.economist.com/specialreports/displaystory.cfm?story_id=14994872 (January 23, 2010).

29. "Getting Warmer," *Economist*; "Scientists Seek Better Way to Do Climate Report," *Times of India*, February 11, 2010.

30. Hansen 2009a.

31. Fukamachi 2010.

32. Trenberth and Fasullo 2010.

33. Oppenheimer 2010.

34. Oppenheimer 2010.

35. Dr. James Hansen, an internationally acclaimed NASA climate scientist, explained the complex nature of the relationship between global warming and rise in the sea temperature.

36. Nicholas Stern, well-known British economist alleged to have said that sceptics are just recycling the old arguments, but the leaked e-mails, argued one commentator that it could potentially provide a lever, but did not consider that in the end it's going to affect the talks in Copenhagen. For example, Bjorn Lumberg, a well-known Danish economist, political scientist, who has been called a sceptic said that there's plenty of science, sufficient science to drive some policy to start reigning in these emissions, as asserted by the IPCC. http://www.abc.net.au/lateline/content/2009/s2767053.htm (December 21, 2009).

37. "Climate Summit Opens in Copenhagen," *ABC*, December 7, 2009, http://www.abc.net.au/lateline/content/2008/s2764519.htm.

38. J. Hewett, "Climate Denier Lords It over Scientists," *The Australian*, January 26, 2010.

39. "Science Damaged by Climate Row Says NASA Chief Cicerone," February 20, 2010, http://news.bbc.co.uk/2/hi/science/nature/8525879.stm; "UN to Review Its Top Climate Panel IPCC," *Times of India*, March 11, 2010.

40. United Nations 2007: 65; Giles 2007: 14.

41. Stern 2009.

42. General A. C. Zinny; the report was prepared by CNA (2007).

43. "Historical Determinism," *Economist*, March 14, 2009. Climate historians in the Lamont-Doherty Earth Observatory of the Columbia University's research confirmed this.

44. "The Clouds of Unknowing," *Economist*, March 20, 2010.

45. Sourced from *ABC News—Special Events Blog* (Margot O'Neill), November 12, 2009, http://blogs.abc.net/au/events/2009/11/conspiracies-and-the-ipcc.html.

46. "Conspiracies and the IPCC," *ABC News—Special Events Blog*, November 12, 2009, http:/blogs.abc.net.au/events/2009/11/conspiracies-and-the-ipcc.html.

47. "Now, Bangladesh Disputes IPCC's Flooding Threat," *Times of India*, April 24, 2010.

48. World Economic Forum, "Will the Environment Lose Out to the Economy?" http://www.weforum.org/en/knowledge/Events/2009/AnnualMeeting/KN_SESS_SUMM (accessed February 23, 2010).

49. "War Footing Needed," *Lateline*, December 8, 2009. His comments followed when it was revealed that the WMO found that Australia's third hottest year on record

was 2009 and that above-normal temperatures were recorded in all continents. Furthermore, the 1990s was the warmest decade for the entire planet since temperature measurement began in 1880. Also, globally, 2009 is the fifth-warmest year on record. But more important, the decade from 2000 to 2009 was warmer than the 1990s, which was, in turn, warmer than the 1980s.

50. "Climate Change Scepticism Growing in US," *Lateline*, December 9, 2009, http://abc.net.au/lateline/content/2009/s2767038.htm.

51. BBC World Service with the Wellcome collection organized philosopher and public intellectual interviewing the chair of the IPCC. The chair confirmed that the IPCC comprised four thousand of the world's climate change scientists; http://www.bbc.co.uk/programmes/p0053j3y (accessed December 30, 2009).

52. A recent example is the discredited work by Andrew Wakefield, a physician, the lead author of a 1998 study published in *The Lancet*, which reported bowel symptoms diagnosed with autism spectrum disorders and linked them to the measles, mumps, and rubella vaccine. Subsequent investigations failed to confirm or reproduce Wakefield's findings. In 2004, the British General Medical Council launched an inquiry into the allegation of misconduct against the physician. On January 28, 2010, it ruled that he acted "dishonestly and irresponsibly." Following the announcement, *The Lancet* retracted his 1998 publication (*BBC World News*, February 2, 2010).

53. http://blogs.abc.net.au/events/2009/11/conspiracies-and-the-ipcc.html (accessed December 21, 2009). The blog was posted by Margot O'Neill on November 12, 2009, at the Copenhagen Climate Change Conference.

54. They released a "State of the Climate" report on March 15, 2010. Also, see "Bureau of Meteorology Backs Up Climate Science Data," http://www.radioaustralia.net.au/pacbeat/stories/201003/s2846811.htm (accessed March 16, 2010), and "CSIRO Backs Global Warming Science," http://www.abc.net.au/lateline/content/2010/s2846667.htm (accessed March 16, 2010).

55. Akbar 2010.

56. Borenstein 2009.

57. Louka 2008: 412.

58. "Climate Change Action Grinds to a Halt," http://www.sbs.com.au/news/article/1162552/Climate-change-action-grinds-to-a-halt (accessed January 11, 2010).

59. Lomborg 2010.

60. http://blogs.abc.net.au/events/2009/11/the-psychology-of-climate-change.html (accessed December 21, 2009).

61. *Economist*, 2010, http://www.economist.com/specialreports/displaystory.cfm?story_id=14994872 (accessed January 23, 2010).

62. "Climate Change Scepticism Growing in the US," *Lateline*, http://www.abc.net.au/lateline/content/2009/s2767038.htm (accessed December 21, 2009). See also the BBC on climate change issue, "Why Is the World Full of Climate Change Deniers?" March 13, 2009.

63. Solomon et al. 2009.

64. "Copenhagen Climate Change Summit Kicks Off," ABC, December 12, 2009, http://www.abc.net.au/7.30/content/2009/s2765603.htm (December 21, 2009).

65. Following the accusation by the climate skeptics about the work of the Climate Research Unit of the East Anglia University, relevant agencies including the British government appointed three independent committees to investigate whether there was any wrongdoing by the unit. Two reports are now publicly available, one by a parliamentary committee and the other by the university-appointed committee under the leadership of Lord Oxburgh. Both reports exonerated the unit from any deliberate scientific malpractice but criticized it on some procedural and management practices, which did not materially affect their research findings. Also see *Times of India*, April 15, 2010.

66. "Scientist in Climate Change Data Row Steps Down," BBC, December 1, 2009, http://news.bbc.co.uk/2/hi/uk_news/england/norfolk/8389727.stm.

67. "Climate Change: Mail-Storm," *Economist*, November 28, 2009.

68. Webster and Franklin 2010.

69. "Getting Warmer," *Economist*, December 3, 2009, http://www.economist.com/specialreports/displaystory.cfm?story_id=14994872; "A Heated Debate: Why Political Orthodoxy Must Not Silence Scientific Argument," *Economist*, November 28, 2009.

70. "Beijing Won't Be 'Bullied' on Climate," *New York Times*, January 31, 2010, published by the *Times of India*, February 1, 2010, http://www.indiaclimateportal.org/Beijing-won-t-be-bulliedon-clin.

71. Goldstone (2010) includes conclusions of few other researched articles in support of his thesis. Interested readers may consult those. R. E. Cincotta and D. Anastasion, *The Security Demographic* (Washington, DC: Population Action International, 2003), and E. Leahy, *The Shape of Things to Come* (Washington, DC: Population Action International, 2007).

72. "5-Day Davos Forum Ends on Note of Humility," http://news.yahoo.com/s/ap/20100131/ap_on__bi_ge/davos_forum (accessed February 1, 2010).

73. McMahon, Parker, and Miller 2010.

74. Scott Bevan, December 5, 2009, "Fears Climate Change Bringing Health Risks," http://www.abc.net.au/lateline/content/2008/s2762793.htm.

Chapter Two

Climate Change and Freshwater

Endowment of natural resources varies between continents, countries and within countries; some are luckier than others. Water is the only natural resource for which there is no substitute, and no life can exist without it. Water also has emotional and identity attachments not typical of other natural resources.

Only an extremely small portion of the 1.36 billion m^3 of global freshwater is available for use and much of it remains buried too far underground to be exploitable. Globally, each year 500,000 km^3 are lost as moisture into the atmosphere, 86 percent from the ocean and 14 percent from the land. The earth's supply of freshwater increases by small quantities, to reach 40,000 km^3—the product of a complex balancing of evaporation and rainfall between sea and land.[1] Along with atmospheric, hydrological, and climatic factors, a host of other variables affect the availability of freshwater.[2]

With rapid global population increases, the need to produce additional food and other lifestyle factors, per capita global freshwater availability declined substantially by the end of the twentieth century. During that century, the world's population tripled, while water use for human purposes multiplied by about sevenfold. Experts further estimate that water use will increase by about 32 percent during 2000–2025;[3] and about one half of the global population will live under water stress by 2025.

South Asia has the second-lowest per capita volume of freshwater in the world; besides the Middle East and North Africa (table 2.1). Shifting patterns of precipitation and runoff associated with climate change will further exacerbate the crisis.[4]

In a declining resource environment, overcoming the water crisis becomes a policy priority for water-deficient countries. India receives little more than 34 percent of her freshwater from outside the country. Perennial riparian rivers

Table 2.1. Annual renewable water resources

Country	Year of Estimate	m^3/yr
Afghanistan	1997	65.0
Bangladesh	1999	1,210.6
Bhutan	1987	95.0
India	1999	1,907.8
Nepal	1999	210.2
Pakistan	2003	233.8
China	1999	2,829.6

Source: Gleick et al. 2006: 225.
Note: The estimates consist of natural renewable surface and groundwater, and flows from other sources; and the data comes from different sources; hence any comparisons between countries should be made cautiously.

provide this water. In this context good policy demands a cooperative water-sharing arrangement with the upper riparian countries. But legacies of distrust, bitterness, and disinclination to cooperate in the subcontinent restrict any capacity to elect such a policy option.[5]

There are few instances where cooperative approach has led to peaceful resolution and success of a water-sharing arrangement. The Danube River water-sharing arrangement is a rare example where seventeen countries share its waters.[6] The Danube success may be attributed to Europe's history of cooperative approach in using their scarce natural resources.

Besides rainfall, freshwater is available from glaciers, mountain lakes, springs, and aquifers. India's constantly flowing rivers (the Brahmaputra, the Indus, and the Ganges) receive water from monsoonal rains, and from the melting snow of the Himalayas. Many countries share the catchment areas of each of these rivers (table 2.2). There are disputes between countries over which country owns a few of the river basins. Nonperennial rivers in the subcontinent receive their waters mostly from monsoonal rain, lakes, and springs.

India's yearly rainfall is limited to three to four months (except in small parts of the south). The annual evapotranspiration rate (in millimeters) is between 1,400 mm and 1,800 mm; while in Bangladesh it is 1,250 mm, and in Pakistan it varies from 1,300 mm in northern Punjab to 2,800 mm in the Sindh.[7]

Many argue that in a water-stressed and water-scarce situation, the reason for conflict or war will be water, not oil.[8] However, in a complex political environment where besides India, two other water-sharing countries have a nuclear capability, the possibility of an all-out arms conflict is a matter of conjecture.[9] Although this remains a remote possibility, scattered water-related skirmishes or fractured approaches to resolve other policy issues

Table 2.2. Major river basins in mainland subcontinental countries and China

Basin	*Area in km²*	*Share of the basin areas in percentages*							
Country name		Afghanistan	Bangladesh	Bhutan	China	India	Myanmar	Nepal	Pakistan
Country area in km2 ('000)		641.9	138.5	39.9	9,938.9	3,089.3	669.8	147.3	877.8
Fenny	2,800	—	34.12	—	—	65.88	—	—	—
Ganges-Brahma-putra-Meghna	1,675,700	—	7.36	2.38	19.12	62.21 [i]	0.13	8.79	—
Indus	108,600	6.68	—	—	11.11 [ii]	26.13 [iii]	—	—	56.09
Irrawaddy	404,100	—	—	—	4.6	3.81 [iv]	91.15	—	—
Kalandan	30,500	—	—	—	—	24.24	74.39	—	—
Karnafuli	15,000	—	74.78	—	—	24.98	—	—	—

Source: Gleick 2000: 220; Wolf et al. 1999: 387–427.
Note: In the table, countries claiming a further share of the basins are (i) China (4.07 percent), (ii) India (0.89 percent), (iii) China (0.15 percent), and (iv) China (0.29 percent).

cannot be entirely ruled out. In fact this has already been happening in the Indian subcontinent.

Although countries may not enter into major or overt conflicts on water-sharing issues, international relations considerations often override technical issues in reaching an optimal outcome on settling disputes.[10] The intrigues of international politics are complex. Often they preclude resolution of water-related conflicts and contribute in protracted conflicts, which sometimes bring changes in existing political configurations.[11] Further, it is undeniable that contentious issues often remain unspoken and influence the outcome of other issues such as water conflicts.

Individuals perform intricate and imaginative maneuvers to achieve political and other objectives.[12] Available evidence suggests that disputes in resource-sharing arrangements may not always directly cause conflict or war, but they create a context in which a country's vulnerability to armed conflict is increased, and shape the opportunities available to would-be belligerents.[13] In the modern era open conflict about water-sharing arrangements has been rare, but indirectly it has been the contributing factor in many conflicts and wars.[14] As freshwater gets scarcer the possibility of conflict increases; and it becomes more difficult to resolve other existing conflicts.

A clear example is the relationship between India and Pakistan, where they always accuse or oppose each other even on trifling issues. Besides, sharing the Brahmaputra water issue may remain contentious in settling ongoing disputes between India and China. Notwithstanding the Indus Waters Treaty of September 19, 1960, both countries have started squabbling over sharing water, as each country proposes building new dams to harvest additional water from the Indus River system. Pakistan has directly linked this issue with bilateral discussions to improve overall relations with India. India and Bangladesh have an agreement on sharing water from the Ganges River, but relations with Bangladesh often soured when the Bangladesh Nationalist Party (BNP) was in power.

Global undercurrents of political events indicate that some countries are already or nearly at war to protect and preserve their water interests (such as the conflict between Botswana and Namibia over the Okavango River). Another example is the Turkish opposition to Kurdish independence which many consider is, in part, due to the Kurds living in the mountains—the location of substantial water resource. Some even believe that Syria has shifted its position from supporting an independent Kurdistan fearful of Kurds controlling its water supply. Conflict between Israel, Jordan, and other Middle East countries over water are well established.[15] Political observers have long argued that Israel's determination to maintain its control of the Occupied

Territories partially related to the availability of water in the territory.[16] A significant portion of Israel's current water supply originates in the occupied West Bank.[17]

HISTORICAL PERSPECTIVE

There were 165 water-related conflicts between 3000 BC and 2006 AD, and 117 of these took place between 1945 and 2006.[18] In the last fifty years, only thirty-seven disputes involved violence, and thirty of those occurred between Israel and one of its neighbors.[19] Definitional differences may be responsible for such a wide divergence in the identification of a number of conflicts.

One possible reason for the increase in the number of conflicts during the latter half of the twentieth century was the rise in terrorist activities to settle conflict. Of the identified water-related conflicts, 23 percent relate to terrorism, as opposed to military targets, tools, or goals. This has significant security future implications for all South Asian countries.

Broadly speaking, water conflict is more of a modern-day phenomenon, indicating that, as demand for water increases with the rapid increase in population and developmental activities, so too will the conflict arising from the desire to control water. Even at village level, such possibilities exist, as was found in a south Indian state. There, the people did not hesitate to divert water through their own field channel network, although they did not require the water. They did it to ensure that the people in the next village could not exert pressure on the irrigation department for more water.[20]

A country's inadequate governance environment and corrupt administration may also contribute in water-related conflicts, where political parties and unscrupulous politicians keep such conflicts deliberately alive. This enables them to offer excuses for their policy failures.[21] For example, Yemen, one of the driest and poorest countries on earth, subsidizes water production cost to produce a social (recreational) drug (locally known as "qat"), the production of which requires enormous amounts of water. Yemen's water crisis is so acute that the United Nations estimates that the country will need to relocate its capital Sana by the end of 2025.[22]

The countries whose inhabitants account for almost 40 percent of the global population share water with other riparian countries. Notwithstanding the existence of an estimated three hundred agreements between various riparian states, the absence of an international protocol on riparian water-sharing arrangements has the potential to make the situation worse, with a deepening climate change–induced water crisis.

SOUTH ASIA'S WATER-SHARING ENVIRONMENT

Domestic and intercountry riparian water-sharing concerns have been ongoing for decades in mainland South Asia. Pakistan's strong commitment to Kashmir is partly due to its belief that Kashmir denies India any opportunity to fully control the water from the Indus River that flows into Pakistan.

Bangladesh and India had a long-standing disagreement on the Ganges River water-sharing issue. The Treaty between the government of the Republic of India and the government of the People's Republic of Bangladesh on sharing the Ganga/Ganges Waters at Farakka was signed on December 12, 1996. Both countries also agreed to use the surplus water from the Brahmaputra River to increase water flow in the lower Ganges, but disagreed on the most effective mechanism to achieve this. Any Brahmaputra water-sharing arrangement might require China's involvement; for technical, hydrological or geological reasons rather than any political reason. Given China's 48 percent ownership of the Brahmaputra Basin at the upper riparian level, it has a natural advantage in regulating the river's downstream water flow. There is, however, no guarantee that China will not use this position to extract maximum political advantage. It is a water-scarce country, where per capita surface and near-surface water is roughly one-fourth the global average and, worse yet, is distributed unevenly.[23]

The Indian media doubted for many years that China had been planning to divert water from the Brahmaputra River to meet its own needs.[24] Recently, China admitted to India for the first time that it was building a hydropower project on the river. The political overtone of China's statement is interesting and needs further examination. The issue is examined from China's internal economic and political perspective; and from China's long-term global interests.

India has bilateral agreements with Nepal on water-sharing arrangements. These include the Koshi Agreement signed on April 25, 1954, and revised on December 19, 1996, the Mahakali Treaty signed in February 1996, and agreements covering the Sapta Koshi and Burhi Gandaki rivers. Anti-Indian feeling has been high in Nepal for some time, and has increased in recent years. The 1990 Act that required two-thirds parliamentary approval for all agreements with foreign countries concerning Nepal's natural resources (including water) is a manifestation of such anti-Indian feeling. Given that all Nepalese rivers feed the Ganges River, it may be that friendship between China and Nepal leads to some water-diversion projects to meet China's thirst for freshwater. The strength in this relationship was evident when Nepal insisted that China be allowed observer status at the South Asian Association for Regional Cooperation (SAARC), a move opposed by India.

Although Maoists currently hold no critical governmental posts in Nepal, they remain the largest single political party in the Nepal parliament, and they strongly influence government policies toward India. Maoists consider India a hegemon country.[25] A recent example is its successful pressure on the government to cancel a deal with the Indian government for printing machine–readable passports from a government security press. This stance was taken notwithstanding the government's failure to meet the deadline imposed by the International Civil Aviation Organization (ICAO) asking Nepal to stop issuing handwritten passports.[26]

Global freshwater resources are unevenly and irregularly distributed. Political borders rarely coincide with watershed borders. Forty percent of the world's population rely on shared water, and a very large section of those affected reside in South Asia. The Indus and the Brahmaputra rivers are considered sources of potential conflict. The way India was divided by ignoring the physical and other related issues reflects a failure both by the leadership of the retreating colonial power and those governing in South Asia since 1947.

Of more than 260 international river basins, only five or more nations share thirteen rivers.[27] The Ganges-Brahmaputra is one such basin, covering an area of 1,600,400 km^2.[28] This basin remains one of the most contentious as the water crisis deepens, where the world's two most populous countries—India and China—along with other smaller countries vie for access to its waters.

Meanwhile, the Mekong River in Southeast Asia has already become a source of major conflict between China, the economic and political superpower in the region, and the economically and politically weaker lower riparian countries. China did not join the Mekong River Commission (MRC) at the commission's formation in 1995. Rather, it chose to remain an observer. This enabled China to pursue independent policies concerning the use of water from the Mekong River.[29] By then China had constructed two dams upstream and was believed to be considering building two more dams. These dams restricted the free flow of water that was vital for the economic survival of the lower riparian countries, affecting such industries as transport and fishing.

In April 2010, a conference focusing on the Mekong River was held in Bangkok to discuss the declining downstream water flow, which was causing a crisis in Thailand's fishing industry. The countries downstream from China blamed their larger neighbor for the reduced water flow, but China is alleged to have declined to accept blame, arguing that any reduced flow was due to climate change and shifting rainfall patterns. There is no guarantee that a similar situation will not arise in future with respect to the use of water from the Brahmaputra River.

Water-sharing arrangements are also a problem in another region—the Nile valley in northern Africa. Former UN Secretary-General Boutros

Boutros-Ghali (when he was Egypt's foreign minister) forecast that the next war in the region would result from conflict over the Nile waters, not from political disputes.[30] Egypt has the major share of the Nile water, although it is one of nine impoverished countries in the entire basin. By historical accident it was able to claim this share in 1929, as Britain (the colonial power at the time) had a vested interest in maintaining the economic and political dominance of Egypt, vis-à-vis the other eight countries in the Nile basin.

During the Cold War, and even after Israel's win in the 1967 Six-Day War, Egypt remained within the US alliance. Egypt's superior military power vis-à-vis other countries in the Nile basin, and its strategic importance in protecting a vital navigational line between the West and the East, also helped it maintain this advantage despite vocal opposition from other basin countries.[31]

The type of power-based relationship always determines, directly or indirectly, the conflict resolution mechanism, as can be seen in the Israel-Palestinian conflict or in US-Mexico relations. In the riparian water conflict between the superpower and a middle-level developing nation, it is simply a case of vast disparity in the military and economic might of both countries.[32] Usually such relations determine the ground rules for dispute resolution mechanisms. There are a few exceptions though, mostly due to historical antecedents, such as in India-Pakistan or India-Bangladesh relations.

India's domestic water-sharing environment is already critical for two reasons. First, unsustainable groundwater extraction has significantly reduced groundwater level. Consequently, groundwater stock has been declining rapidly, as has occurred in Quetta, Pakistan. Observers in the subcontinent generally concur that conflicts were becoming endemic at all levels in India.[33] India's 1987 and 2002 national water policies are inadequate to meet any current, let alone emerging, challenges.[34] These two policies can be considered as being in name only, as neither provided any indication that policymakers understood the critical importance of water to India's development efforts and unity.

In an interactive workshop held during the January 27–31, 2010, annual meeting of the World Economic Forum in Davos, Switzerland, the attendees virtually agreed that the issue of water is intrinsically linked to politics, but that it only entered the policy stage by the back door. The participating politicians highlighted the practical difficulties and complications of adopting a universally acceptable policy to confront the urgent need to address water depletion. They were almost all of the view that the rapidly deteriorating water environment could lead to economic collapse.[35]

The World Bank report argued that the ministers either do not understand or are not courageous enough to adopt policies that might achieve positive policy outcomes. It became clear from the statements of the Finance Minister and the Minister of Water Resources. The former warned of a growing set of minor civil wars over water, while the latter wryly noted that he was really the "Minister of Water Conflicts."[36] The country's federal structure and poor political leadership has largely, yet gradually, contributed to this deterioration over the years. The situation is now so politically complex that any resolution of conflict is virtually impossible, and any bold attempt to try to do so may threaten India's unity.

In South Asia water became a weapon of war following the 1947 partition of India. The Indus River water-sharing dispute in the northwest and the Farakka Barrage dam dispute in the east are two of the most publicized and contested disputes. The Indus River dispute was resolved in 1960, following mediation by the World Bank.

The Farakka proposal was nearly a century old by 1947. A series of formal investigations recommended in favor of building the dam to maintain the navigability of the Kolkata port. Following partition, the India government started building the barrage to keep the port's navigational channel open. Pakistan opposed the dam on the grounds that it would divert water from the Ganges River to India at the expense of the needs of eastern Pakistan. In 1976, Maulana Abdul Hamid Khan Bhashani, a political leader within Bangladesh, began a long march from that country to India to halt construction; yet the Bangladesh government did not allow the march to proceed beyond its international border. The Farakka issue has become a part of the Bangladeshi psyche ever since.[37]

There have been a few other water-related conflicts in South Asia. There have been military and terrorist action to destroy dams, cut off supplies or pollute water sources, but these form part of wider campaigns to disrupt economic life or pursue a political cause.[38] In 2002 Maoists in Nepal's west destroyed microhydro projects and damaged water supplies, and in 2000 Chinese officials in Guangdong Province blew up water channels to prevent a neighboring country from diverting water.

Pakistan was established in 1947 and its eastern part seceded in 1971 to become Bangladesh. Both countries bear a common Islamic identity and have a history of long-ruling military or military-backed governments.[39] The difference between the Islamic nonsecular orientation of these two states compared with India's secular and democratic orientation has often stood in the way of a beneficial cooperative approach to resolving contentious water-sharing issues. This situation did not develop overnight; the roots of the problem developed long before partition.[40]

The reasons for conflict have not always stayed the same in the past, nor will they remain so in future. As situations change at ground level, so will the reasons for conflict. Events during the last few decades clearly suggest that resource control issues will dominate the approach by many nations to the security environment,[41] and access to freshwater will be one such important resource control issue. The ability of modern technology to dam and divert rivers, irrespective of any drawbacks, has increased the likelihood for tension.[42] Many conflicts in the African continent owe their origin to the control of critical resources, such as in drought-ravaged Darfur, Sudan, and Huila, a southern province in Angola.

Riparian countries need to remain aware of any direct and indirect strategic concerns and related security issues that might emanate from water scarcity, climate change, and environmental degradation. The greenhouse effect of climate change is the archetypal global problem of the present era and will be most evident in water-related problems, some of which will be local rather than regional.[43] The countries of South Asia need to realize that the greenhouse effect has the potential to change rainfall patterns, thereby accelerating conflicts between communities. Other types of conflicts affected by increased water scarcity include disputes between large and small industries, intrapolitical jurisdiction differences, and the private versus public struggle for control of water.[44]

Bangladesh, India, and Pakistan, as the three major countries of the subcontinent, all receive a significant proportion of their freshwater from outside their respective borders. Of these three countries, Bangladesh is virtually dependent on other countries for most of its renewable water resources (table 2.3), followed by Pakistan and India. Bhutan and Nepal are largely independent of other countries for their renewable freshwater needs. Despite being a water-scarce country, China has minimal dependence on water from outside

Table 2.3. Water dependency, groundwater withdrawal, and desalinization capacity of mainland South Asian countries and China

Country	*Percentage of renewable water originating outside country in 2000*	*Annual groundwater withdrawals (m^3 per person) in 2000*	*Desalinization capacity (m^3 per day) in 2005*
Afghanistan	15	—	—
Bangladesh	91	98	—
Bhutan	0	—	—
India	34	223	115,509
Nepal	6	—	—
Pakistan	76	490	189,250*
China	1	47	143,000

Source: Columns 2 and 3 (Clarke and King 2006: 94–99); column 4 (Pacific Institute, http://www.worldwater.org/table16.html). Gleick et al. 2006: 311–13.

*In 2005 Pakistan had two desalinization plants (Karachi and Gwadar), while China had one (Yantai)—all three are among the one hundred largest global plants. The total capacity of Pakistan's plants is 189,250 m3 per day, while the total capacity of the Chinese plant is 143,000 m3 per day.

due to the fact that most of Asia's mighty rivers originate in Chinese territory. In the absence of a universally accepted protocol or legal framework for sharing riparian waters, China will need to deal with lower riparian countries through which flow some of Asia's rivers. Many consider that China has shown less than acceptable standards of concern for some of these smaller countries.

Historical issues that have driven freshwater policies in many riparian countries are both illuminating and interesting:

- Conflicts are largely determined by the power relations between the parties involved (such as relations between the United States and Mexico on the subject of water-sharing arrangements);[45]
- Where the dominant state lies upstream, it holds all the cards. On the other hand, where the more powerful state is located downstream, it can compensate for any geographical disadvantage by military means and may even become a trigger for military conflict (such as Egypt's policy toward the upstream countries of the Nile); and
- Where governments include the diplomatic fallout of water projects in the equation, other political conflicts often also determine the outcome of water distribution conflicts.[46]

Bangladesh, India, and Pakistan are at various stages of becoming water-scarce countries (table 2.4). By 2050, these three countries and Afghanistan will truly become water-scarce countries. It could lead to a fully fledged crisis, with severe impacts on food production, health, nutrition, and the environment. Overcoming such a situation will not be easy. One thing is certain; it is now time to begin a fundamental reform of the water sector.[47]

Table 2.4. Internal renewable water resources of the mainland subcontinental countries and China: 2000 and 2050

Country	*Cubic meter per person per year in 2000*	*Cubic meter per person per year in 2050*	*Percentage decline per person 2000–2050*
Afghanistan	2,527	761	70
Bangladesh	764	396	48
Bhutan	45,561	17,059	63
India	1,249	802	36
Nepal	8,601	3,781	56
Pakistan	1,756	721	59
China	2,194	1,924	12

Source: Food and Agricultural Organization Aquastat 2003 (Clarke and King 2006: 94–99).

Bhutan and Nepal will have plenty of freshwater from their own sources, but this does not imply that their agriculture will be free of drought, as rain may neither fall at the right time of the agricultural cycle, nor in the right amount. Also, receding glaciers may no longer be able to provide the current level of water supply.

The United Nation's Food and Agriculture Organization (FAO) forecasts make it clear that China occupies a major position in drafting any subcontinent freshwater management policy. China, with about 22 percent of the world's population, has access to only about 8 percent of global freshwater and it is facing severe water shortages.[48] Distribution of freshwater is uneven in China: the parched northeast China has only one-fifth of the per capita water that is available to south China; northern China has highly variable rainfall and is susceptible to dust storms and soil erosion; and south China, although being wet, has heavy rainstorms that cause erosion.[49] A number of media reports further confirm that the situation in parts of China is becoming increasingly critical.

From India's perspective, policy environment becomes complex, as China is its competitor—economically and politically—in the region and globally. China is one of the three countries who voted against the draft 1997 UN Convention on the Law of the Non-navigational Uses of International Watercourses. While it is politically difficult for a country to make a unilateral decision on managing riparian waters without giving at least some ground to the needs of lower riparian countries, it is possible to push for arrangements that remain favorable to the upper riparian country during negotiations, particularly if that country is economically, politically, and militarily powerful. In this context China's relations with the smaller regional countries and with Pakistan becomes a strategic foreign policy concern for India.

For smaller countries in South Asia the experience of Mexico—a lower riparian country to the powerful United States—provides an example of how difficult such relations can be in certain circumstances. Water and security problems will not vanish; however, the intricacies of international relations often directly or indirectly force countries to adopt policies that defy logic.[50]

Besides water-sharing arrangements, the other critical issue is the quality of the riparian water. For example, while the United States complies with its legal requirement to maintain a minimum flow of water from the Colorado River to Mexico, it has minimal concern for the heavily salinated water.[51] At international levels it is accepted that water and environmental issues are intertwined.[52] This is increasingly becoming a critical issue globally, as Executive Director of the United Nations Environment Programme (UNEP) Achim Steiner wrote in his Foreword to the Pacific Institute's 2010 report "Clearing the Waters: A Focus on Water Quality Solutions": "The challenge of water in the 21st century is one of both quantity and quality. . . . Contaminated river systems, coastal waters and other ecosystems are not only a health risk, they are also a risk to livelihood and economies."[53]

Pollution levels in rivers, lakes, and other water sources are high in subcontinental countries and China. Aside from general contamination of water resources, the greatest danger comes from the constant introduction of existing and new chemicals. For example, about seven hundred new chemicals are introduced into commerce each year in the United States alone.[54] Another study estimated that the value of ecosystem services is double the gross national product of the global economy, and the role of freshwater ecosystems in purifying water and assimilating wastes has been valued at US$400 billion.[55]

New issues continue to emerge from changes caused by climate change, in human activities and behavior, in direct and indirect political reconfiguration, and in the global economy. The freshwater crisis in northwestern and eastern regions of South Asia is largely a product of political decisions dating from the creation of Pakistan and Bangladesh. The fluid and transspatial qualities of water streams mean they play a crucial role in interstate conflict. Rivers and tributaries present a clear contradiction between the natural borders of ecoregions and the political border of nation-states. Water links ecological regions that lie in different political domains.[56]

In addition to surface water, a major source of freshwater is held in aquifers. Naturally occurring arsenic pollution has heavily contaminated Bangladesh and parts of the Indian and Chinese aquifers. The origin of this arsenic pollution is geological.[57] Arsenic affects humans through air, food, and water, and arsenic poisoning has affected about half of Bangladesh's population. This country's experience forced the World Health Organization (WHO) to reduce its earlier allowable arsenic level from 0.05 mg/liter to below 0.01 mg/liter. In addition, mineral contamination of groundwater in India is also significant: about 175 of the 593 districts in India fall into this category.[58] While small doses of some minerals for a short time may not be injurious to human health, long-term exposure often leads to disastrous health consequences. For example, fifty years of spraying the pesticide DDT has resulted in contaminated groundwater in many Indian states.

In the mid-1990s the then US Congress-based Office of Technology Assessment detected the existence of over 175 organic and 50 inorganic chemicals among which are liver and kidney toxicants.[59] Globally, it is estimated that groundwater as deep as 2.8 km may have rich bacterial flora.[60] As the subcontinental populations are largely dependent on aquamarine water, this becomes both a major health and economic concern. First, it would require intensive research to establish what else one needs to know before reducing future dependence on aquifers.[61] Second, funds required for both research and infrastructure development before these waters are available for human and related economic purposes will be beyond the financial capacity of most levels of governments. This is a major strategic concern for all subcontinental countries.

As Bangladesh, India, and Pakistan significantly depend on water flowing from upper riparian countries, they have to remain continuously vigilant about water pollution in upper riparian countries. A European example occurred on November 1, 1986, when a fire at a Sandoz-owned chemical factory near Basel, Switzerland, sent tons of toxic chemicals into the nearby Rhine River and turned it red. The Rhine River flows through Switzerland, Germany, France, and Holland before flowing into the North Sea, and the Dutch were particularly concerned about the presence of mercury in the chemicals.

Of the three constantly flowing rivers in India, both the Indus and the Brahmaputra originate in China and flow through India, Bangladesh, and Pakistan before entering the sea.

China has been a major polluter of many rivers. In China, 80 percent of the major rivers are so degraded that they no longer support fish.[62] China discharged 40 million tons of industrial waste and raw sewage into the Yangtze River every day during the 1990s.[63] The country also emits the largest amount of organic water pollutants. In 1998, China was emitting 7 m kg per day, compared to the United States' 2.5 m kg, India's 1.7 m kg, Russia's 1.5 m kg and Japan's 1.4 m kg. However, if we consider per worker/per day emissions, the difference between China, India, and Pakistan was not that large—at least in 2000. In China, it was 21 kg, followed by India (19 kg) and Pakistan (18 kg). Nepal and Bangladesh both recorded 14 kg per worker.[64]

Outside China, India is home to the most polluted waters in Asia. This includes both the perennial and rain-fed rivers. The Ganges River, while not originating in China, remains one of the most polluted in the world. Its pollution level has worsened despite spending hundreds of millions of dollars. In 1994, the percentage of violation of the pollution standard was 80 percent, which increased to 100 percent in 1999.[65] India's Supreme Court recently expressed its displeasure with the misuse of the Ganga Action Plan funds launched in 1985.[66] India's Supreme Court in a judgment in 2009 stated that polluters must pay, and imposed a heavy fine on a polluting company, arguing that reducing environmental pollution was a priority in view of the global mandate to reduce carbon emissions.[67] Unfortunately, this situation remains unchanged and the rivers show no visible sign of improvement.

The condition of the Indus River is also poor. Before it enters the Arabian Sea in the port city of Karachi it is virtually just a large sewer. The Indus River now often fills only a small fraction of the riverbed. It used to be 5 km wide near Karachi, it is now only 200 m wide, and fishing has disappeared, seawater has invaded agricultural land, and water for domestic use is desperately short.[68] Pakistan's agriculture is largely dependent on the Indus River system, as are the northern regions of India. The Indus irrigation system is the world's largest contiguous irrigation system. If such levels of pollution occur in the Brahmaputra, the situation will be disastrous for both India and Bangladesh.

Aside from the economic, aesthetic, and ecological importance of maintaining clear and unpolluted water flows in rivers, another concern is that 1 m^3 of contaminated wastewater spoils up to 10 m^3 of pure water. This is a critical issue in most subcontinental countries where there exists a huge shortage of uncontaminated freshwater.

The situation is complex for India because it becomes both a victim and predator at the same time. It becomes a victim if China's pollution levels affect the waters from the Brahmaputra and the Indus rivers. As both rivers flow into Bangladesh and Pakistan respectively, they can then accuse India of not doing enough to stop polluted water flowing into their countries.

Another major concern is the saline water intrusion in coastal regions, particularly in India,[69] caused by the unsustainable extraction of groundwater from the aquifers without creating appropriate recharging facilities. Also, a sharp fall in aquifer water levels creates twin problems of construction cost increases and waterlogging. Aquifer levels have been falling at an average rate of 2–3 m each year in a number of regions. The resultant depletion of groundwater aquifers has some analysts predicting that 25 percent of India's harvest may be at risk in the coming years.[70] In Gujrat state, the unsustainable extraction of waters in the earlier decades has already caused a rapid reduction in production levels.

In Pakistan, the aquifer levels of the Punjab are dropping by 1 m a year, and in Baluchistan by 3.5 m a year, while the level is also rapidly declining in the Northwest Frontier Province.[71] People from water-scarce provinces in Pakistan have started moving out of many villages. Many activists believe that by 2010, Quetta, the capital of Baluchistan, will have exhausted its water supply.[72]

The largest consumer of freshwater in South Asia is the agriculture sector, primarily for irrigation purposes. Irrigation efficiency range from 25 percent to 40 percent compared to 50–60 percent in Japan, Israel, and Taiwan. The real costs of new irrigation in India doubled between 1970 and 1990 and in Pakistan during 1980–1990. The development of new sources also imposes high environmental and other related costs, such as on human resettlement.[73]

Unsustainable extraction and low water productivity is common in almost all developing countries. For example, Mexico is extracting water from aquifers 50–80 percent faster than the rate of regeneration. While the city is yet to go dry, about one-third of the city's water is now pumped to an area of 2,300 m above sea level, some from as far as 300 km. Mexico will run out of water by around 2015.[74] Public and household cost implications are enormous, as somehow water has to be made available to meet the basic needs of citizens.

With the predicted rise in temperature caused by a CO_2 rise in the atmosphere, a further concern is that the increasing rate of evaporation and further loss of freshwater in the subcontinent will create problems for the most vulnerable groups in the community.[75] The effects of such changes will not be uniform, as these vulnerable groups lack access to funds and technologies that could assist adaptation to climate change.[76]

The IPCC has argued that if global corrective measures were to be taken now, it would cost no more than 0.2 percent of global Gross Domestic Product (GDP), but a failure to do so would cost up to 5 percent of global GDP by 2030.[77] Notwithstanding the debate on who is to blame, it is clear that the situation is deteriorating rapidly. Former British foreign secretary Margaret Beckett succinctly highlighted the critical nature of the issue at the UN Security Council (UNSC) in 2007. She even related the Darfur and Somalia crisis to climate change. She emphasized that changes in the world's weather are set to put millions of dispossessed people on the road, as resources become scarcer.[78]

The accelerated rate of climate change has already led to a dramatic increase in the number of geophysical disasters. An average of 326 climate-related disasters occurred (reported) each year between 2000 and 2004—more than double the levels reported in the early 1980s.[79] During the period 1991–2000, an average of 256 natural disasters occurred per year, with 97 percent of victims coming from developing countries.[80] The intensity of some of the disasters that have occurred in recent years is beyond comprehension.

The International Federation of Red Cross and Red Crescent Societies (IFRC) reported that the number of people killed by natural disasters in 2008 was 235,736, which was surpassed only on December 26, 2004—the day of the Indian Ocean undersea earthquake and resulting tsunami. The estimated damage bill was more than US$181 billion and three-quarters of the devastation was wreaked on developing countries, which suffered 99 percent of the fatalities.[81] The bushfires in Victoria, Australia, during the summer of 2008–2009 were the worst in the country's history: The United States has experienced similar bushfire disasters and also hurricanes such as Katrina in August 2005.

Even without the climate change phenomenon, global freshwater shortages are an unpleasant reality, as are droughts and floods, the intensity of which continues unabated. Weather transformations have caused changes in rainfall patterns; some areas get more rain at the expense of others (e.g., northern Europe is now receiving 40 percent more rainfall compared with southern Europe). Worldwide, the number of significant flooding episodes has increased each decade from six in the 1950s to twenty-six in the 1990s.[82] Future droughts and floods are likely to be more intense and qualitatively different from those in previous years.[83] This is clear from evidence showing that significant changes in plant and animal behavior have also taken place in Europe and other places over the last one hundred years.

In some circumstances, climate change may be of limited benefit:[84] farmers in middle to high-latitude areas would benefit from warmer temperatures, and for them it may be economically beneficial (such as in India's Ladakh region).[85] However, the net economic effect is likely to be negative for countries in low-lying latitudes, away from the equator, where agriculture dominates the economy with poor productivity, outdated technology, degraded land, and poor infrastructure and investment. Such countries will not cope with the damage from climate change.[86]

In the policy arena, a prominent NASA scientist considers that the approach made by politicians does not effectively reduce carbon emissions. He argues that the focus should be to lower consumption of carbon emitting products.[87] Yet lawmakers are afraid to confront fossil fuel business interests. In support, he argues that the fight to save the planet becomes difficult when there are four energy lobbyists to every member of the US Congress. There is little doubt that politicians in the debate have often added fuel to the confusion, as noted in an *Economist* editorial: "The problem lies not with the science itself but with the way the science has been used by politicians to imply certainty when, as often with science, no certainty exists. . . . [Although] most research supports the idea that warming is man-made."[88]

The Copenhagen Climate Change Congress in March 2009[89] concluded that, with the passing of each month, the world is approaching a tipping-point beyond which it will struggle to reverse the effects on the planet.[90] The congress confirmed that

- sea levels are rising twice as fast as the original assessment of scientists;
- over the next one hundred years, the global temperature will rise 4°C–5°C, turning most of southern Europe into a desert; and
- eighty-five percent of the Amazon rain forest is at risk of destruction.

Members at the congress concluded that if present trends continue, sea levels could rise by 80 ft or more—levels unseen in three million years.[91] The global community, generally speaking, is aware of these issues. Even the US Congress confirmed that time was running out for the global community. The UK Meteorological Research scientists also concluded that climate change has already reached a level that, even if the world's best efforts were made now, combating climate change would likely offer no more than a 50:50 chance of keeping temperatures below the threshold of disaster. Yet, there still needs to be agreement on a set of time-bound remedial measures.

Some scientists also believe that if people warm earth, they can probably cool it. Suggested mechanisms to combat this warming include dumping iron into the oceans to promote huge blooms of phytoplankton, solar shade, and injecting sulphate particles into the stratosphere to artificially create cloud formations. There are no as yet proven mechanisms for reducing the pace of global warming, and any suggested mechanisms do not have any compensation for emitting carbon dioxide. All of the engineering schemes designed to achieve planetary heating balance will have some negative and potentially sensitive consequences.[92] Most of these are just ideas, and entrepreneurial scientists are testing the water with the objective of securing funding for such research projects.[93]

Irrespective of their scientific and economic viability, two important concerns remain: first, any geoengineering project has to be enormous, thereby causing disruption to ecosystems. Second, the substitution effects of alternative options to minimize the negative effects from climate change could be unmanageable. For example, when the supply of productive land is limited, it may only be possible to establish new forests at the expense of those crops required to produce food to feed increasing populations.[94]

The European Environment Agency (EEA) also concluded that global warming has accelerated in recent years.[95] Some scientists are even arguing that by 2100, there could be warming of the world by 11°C,[96] although the IPCC predicts that it might only rise between 2°C and 4°C. There is little scope to disagree with the underlying message of these findings, as the global scientific community, by nature, tends to be conservative in its thinking.[97] Even the IPCC now acknowledges that it underestimated the reality of climate change, and that global warming was likely to accelerate at a much faster pace and scope, causing widespread environmental damage (including to the polar ice caps).[98]

Scientists have established the reasons for earth's energy imbalance that influence climate change patterns. Preliminary findings of the International Polar Year survey by ten thousand scientists, over a two-year period, revealed new evidence that the ocean around Antarctica has warmed more rapidly than the global average. During the 2007–2008 survey, a special expedition to the Arctic found an unprecedented rate of floating drift ice, providing compelling evidence of changes in the region.[99] Scientists from the National Snow and Ice Data Centre at the University of Colorado in Boulder found that shrinking Arctic ice (and exposure of extra sea to radiation) was warming the world at an accelerating pace; yet a global consensus on cooling the planet remains elusive.[100]

There is further evidence that vast oceans are now heating up to a depth of more than 1,000 ft, confirming that the earth's temperature is rising.[101] Besides rising sea temperatures, scientists have unearthed new evidence of how CO_2 levels impact on ocean life. Upon entering the atmosphere, CO_2 is absorbed by the ocean, causing increased acidity in the water. This concentration prevents some sea creatures from forming thick shells, and has the potential to pose a real threat to the oceanic food chain.[102]

Such developments confirm that the climate change phenomenon is boundary neutral, and that pollution generated by one country inevitably affects other nations through no fault of their own.[103] The critical role of sea temperature in global rainfall and drought is already an established weather phenomenon: the *El Niño* and *La Niña*.

Despite the critical importance of sea temperature in global rainfall and climate, research on sea-linked and related fields has been scarce. Sea levels rose by 20 cm—1.5 mm annually between 1870 and 2001. During the 1990s and up to 2006, it increased at twice that rate, to more than 3 mm per year.

The seas are vital in maintaining ecological balance, as well as being an important source for food, lifesaving medicine and other product.[104] Oceans, rivers, lakes and other water bodies are crucial links in the human food chain. For further discussions on harvesting sea food, refer to the section on food in chapter 4.

VULNERABLE SOUTH ASIA

South Asia has the largest number of poverty-stricken people in the world. South Asia's population density is high; particularly so in Bangladesh. It will be a few decades before the population stabilizes in each of the region's countries.

Rainfall is uneven and occurs only three to four months of the year, except for a few pockets in south India that receive winter rain. Even without the climate change phenomenon, these countries remain vulnerable to erratic climatic patterns, frequent floods and droughts that affect a large number of people every year. For example, in these countries flooding affected forty million people between 2000 and 2004. The 2007 monsoon alone displaced fourteen million people in India and seven million in Bangladesh. During this period globally, on average, climate disaster affected one in nineteen people (or 5.3 percent of the population) in developing countries. The corresponding figure for countries in the Organisation for Economic Co-operation and Development (OECD) was one in fifteen hundred.[105] The situation will worsen with a rising sea level that will displace millions of people.

The economic losses from the great floods of the 1990s are ten times those of the 1960s in real terms. There has been a thirty-seven-fold increase in insured losses since the 1960s. Given the trend toward multiple risk insurance cover, which normally includes flood losses, insurance losses will continue to rise. Unlike citizens of developed countries, those living in South Asia receive little insurance against flood and drought. In future, the loss felt by those in developing countries will be more pronounced and severe.

Climate scientists have identified nine "tipping-points" of climate change that will reach a critical point in the twenty-first century.[106] However, the timing of these elements varies. The Indian monsoon is one of the identified tipping-points;[107] human-induced factors (such as fossil fuel consumption) are another. Among the other tipping-points are

- global climate change (temperature, precipitation, runoff, and sea level rise);
- altered resource availability (water and food, and changes in access to or quality of minerals);
- political disputes (tension and aggression);
- regional conflict (such as the conflict between India and Pakistan on the construction of new dams on the Indus River tributaries);
- the Farakka conflict between India and Bangladesh; and
- global conflict (such as that occurring in the Middle East).[108]

Given that the populous countries of South Asia fall in the water-scarce-stressed group and large sections of their population are engaged in agriculture, national and regional policymakers have no option but to adopt proactive policies to minimize suffering, and must not depend on reactive policies. Further, the intensity of droughts and floods may not always be within the grasp of policymakers. Proactive policies will have to be adopted at different levels and be risk based. This is fundamentally important, as experience has shown that, by adopting proactive policies, developed industrial countries have minimized loss of property and life in disasters.[109]

From South Asia's perspective, balancing freshwater demand with supply, and ensuring genuine participation of stakeholders in policy development is essential.[110] Water is a scarce resource and equal access to it is vital and central to both life and justice. Nature's limits of renewability and the social limits of water equity should govern the patterns of usage and its accessibility; and this is only possible with the active participation from stakeholders in policy development.[111] Such participation implies stakeholders are fully aware of the relevant issues that enable them to comprehend the importance of both the proactive and complex interlinking nature of water policies and water politics.

Politically South Asia remains a volatile and unstable region.[112] Its capacity to face the climate change crisis is limited; yet a regional approach will assist the countries of South Asia to minimize their economic and social costs. Climate change impacts exert extra pressure on these societies and their ecosystems. These may create an environment that an increasingly globalized and interconnected society may be unwilling to tolerate, and one that may unleash a type of violence that will challenge the capacities of weak nation-states.[113]

RECEDING HIMALAYAN GLACIERS

The Himalayan glaciers are receding and gathering less water. This could eventually turn once perennial rivers into seasonal ones. The 33,000 km^2 of glaciers amidst some of the world's highest mountains form the largest concentration of glaciers outside the polar ice cap. Seven of the great rivers of Asia are in this group: the Ganges, Brahmaputra, Indus, Sulween, Mekong, Yangtze, and Huang Ho. About 70 percent of their water remains frozen in glaciers. The Ganges River, lifeline to India and Bangladesh, receives about 70 percent of its water from the tributaries in Nepal which, in turn, receive their water from the Himalayan glaciers. Glaciologist Doug Benn has been studying the mighty Ngozumpa glacier in Nepal for decades and thinks that without a substantial fall in temperature this enormous glacier will vanish.[114]

The IPCC's acknowledgment that its report on the Himalayan glaciers did not follow its own protocol gave ammunition to the climate skeptics. However, in acknowledging its failure, the IPCC was right to argue that it had created a heightened awareness about the real threat to Himalayan glaciers and the thrust of its assertions in its reports.[115]

The Nepal-based International Centre for Integrated Mountain Development (ICIMOD) has been studying the Himalayan region for more than three decades and argues that there are many uncertainties concerning the impact of climate change on the Himalayan glaciers, as temperatures are changing more rapidly at altitude. It further added that while the situation may remain normal for several decades, changes may happen abruptly.[116]

The Indian Space Research Organisation (ISRO) and the Geological Survey of India (GSI) have argued that there is no evidence to prove that the current glacial recession phase is a consequence of climate change, but have expressed concern about the health of the glaciers. The worst is the Suru basin in the Ladakh region of Jammu and Kashmir, where glaciers recorded a 38 percent shrinkage between 1969 and 2004.[117] The Chandra, Bhaga, Parbati, and Warwan basins are the other four recording the worst recessionary trend over the same period. One Indian-based nongovernmental organization (NGO), Navdanya, argued that the government data showed a false complacency. It claimed that the thickness of the Siachen glacier in the eastern Karakoram Range of the Himalayas has retreated by 8 to 10 m since 1995 and, similarly, the Gangotri glacier in Uttarkashi District, Uttarakhand, India, is retreating at the rate of 4 cm a day.[118] A documentary by Greenpeace asserts that in the past thirty years the glacier has retreated at the rate of 30 m every year. In 1976, experts claimed, it was almost a kilometer from where it is today. Vijay Kumar Joshi, a former GSI director, however, disputed the claim.[119]

The current information on the Himalayan glaciers is generating debate, but the available information needs to be reconfirmed by hard data. Still, the people in the region are certain of the change in weather patterns. Farmers meeting at the end of 2009 in the Himalayan town of Rishikesh heard that vegetation was retreating upward.[120]

Irrespective of the time taken by glaciers to melt, ultimately river flows will decline after their initial burst. Recent samples from the Himalayan glaciers confirm that two markers were missing from those usually found in ice cores. These markers are apparently absent because the glacier had lost all ice that had formed at any time since the mid-1940s, rather than just (as was previously known) some ice from its edges. The IPCC estimates that the Himalayan highlands will warm at about twice the average global rate over the next century; and the situation is likely to deteriorate. Estimates of glacial retreat vary from one-third by 2050 to two-thirds by 2100.[121]

The Chinese Academy of Science has also argued that if there is no reversal in the rate of temperature rise, the glaciers on the Qinghai-Tibet plateau will rapidly shrink, from 500,000 km2 in 1995 to perhaps 100,000 km2 by 2030. The only beneficiaries will be the mountain shepherds.[122]

Nepal has some thirty-three hundred glaciers, of which twenty-three hundred have glacial lakes. The United Nations has warned Nepal that glacial lake bursts could trigger huge losses of life and property downstream. Many glaciers are sinking 30 to 60 m each decade, and at least twenty of those glacial lakes are at risk of bursting their banks. Such glacial retreat will create severe problems for Nepal and lower riparian countries.[123] The receding glaciers are also exposing fragile soil to summer sun, causing soil desiccation and erosion of top soil (a valuable natural resource). This is leading to a massive silting of rivers in downstream India and Bangladesh.[124] The situation continues to deteriorate with the increasing rainfall intensity and rising temperatures.

GLACIERS IN THE NON-HIMALAYAN REGION

In Japan, many glaciers have retreated by between 30 and 60 m over the last three decades. Worldwide, other regions are experiencing similar effects. The Athabasca glacier in Alberta has retreated by 1.5 km in the last seventy years and lost 16 m^3 of water each year—more water than the glacier has regained from winter snows in recent years. The Peace River, the largest of the western Rockies' rivers, now has only 60 percent of its mid-twentieth century flow.

The Swiss glaciers sank a meager 1 percent in a twelve-year period to 1985, but lost 18 percent of their area in a fifteen-year period to 2000. Switzerland has now closed two of its ski fields. An OECD study highlighted the plight of small businesses in the Austrian Alps region where banks had been refusing to lend money to some resorts at a lower altitude because of receding snowline in the Alps. Dr. Marcia Phillips, a European permafrost researcher in the Alps noted that the Morteratsch Glacier in the Swiss Alps is receding rapidly, estimating that in one year it receded between 80 and 85 m. The author of the OECD report and other key persons in the region agreed with the researcher.[125] South America is experiencing a similar situation, as receding snow and deforestation expose land to tropical heat and soil desiccation.[126]

In the medium to long term, the receding glaciers may create political conflict between India and China. They are already engaged in border disputes. Melting snow in the mountain ranges will escalate these disputes by determining new boundaries. Europe is facing this problem by starting negotiations with the affected countries to determine new boundary lines. The Italian and Swiss foreign ministers met in March 2009 to discuss the demarcation of a new border, following the melting of snow in the European Alps.

Global warming–induced climate change requires global action, but actions are also required at national and regional levels to address the issues effectively.[127] Without an international framework, countries have to act on their own,[128] lest they continue to suffer the effects of global warming. There are minimal solutions at an individual level,[129] so countries need to initiate policies within their limited capacity to achieve policy objectives that minimize suffering and loss of life and income streams.

SPECIFIC ISSUES OF CONCERN FOR INDIA

India has two major river systems: the Himalayan and the Peninsular. The Indus, Ganges, and Brahmaputra rivers and their tributaries are part of the

Himalayan system.[130] India shares six river basins with seven countries. The number of countries with whom India shares river basins increased during the second half of the last century. China became a basin partner following its annexation of Tibet. India is both an upper and lower riparian country: China, Nepal, and Bhutan are upper riparian countries, while Bangladesh is a lower riparian country. This puts India in a difficult situation, as its approach to water-sharing issues is subject to intensive scrutiny by countries that may have hidden political agendas.

India also faces a perception problem. China and India are economically and politically in the same league, but Nepal, Bhutan, and Bangladesh are not. Nepal and Bhutan are economically poor landlocked countries. Bangladesh has economic, law and order, and environmental problems. Pakistan is in a different league because of its nuclear power status and its extremely volatile political and security environment.

Rainfall in India is limited to three to four months of the year. With an average rainfall of 1,170 mm per annum, India receives about 4,000 billion cubic meters (bm^3) of water per annum. About 1,876 bm^3 of this water is the average annual surface flow in rivers, 80 to 90 percent of which comes in the rainy season. During the remaining months, the gross available quantity is about 400 bm^3.[131] India receives about 34 percent of its renewable freshwater from outside the country. It currently withdraws about 26 percent of the available freshwater. The available yearly per capita freshwater in India is 1,000–2,000 m3, making it water-stressed and water-scarce around 2025.[132]

India is extracting groundwater at an unsustainable rate. The situation in Maharashtra in western India has already reached a critical level, as the sugarcane farmers use water at an unsustainable rate. In Punjab, the water table is dropping by 1 m every year. In the coastal regions of Tamil Nadu, Andhra Pradesh, and Gujarat overextraction has allowed salt water into aquifers.

The annual evapotranspiration rate varies between 1,400 mm and 1,800 mm. It is highest in west Rajasthan, parts of Karnataka, Andhra Pradesh, and Tamil Nadu; and, in some parts, exceeds 1,800 mm. It is greater than the 1,400 mm in coastal Karnataka (a narrow belt adjoining the Himalayan ranges running through Punjab to West Bengal), parts of the Bihar plateau, eastern Uttar Pradesh, and Assam.

India had a population of 1,008.9 million in 2000. This number will increase to about 1,528.9 million by 2050. In percentage terms, this increase is the lowest in the subcontinent, but, given India's large population base, the increase in absolute numbers is among the highest in the world. India's 2005–2006 fertility rate was 2.7.[133] Ten states have achieved replacement level and few others are close to attaining that level. With increasing population and development activities, states are constantly exploring additional sources of freshwater. There is an increase in the intensity of existing water-sharing conflicts, and an eruption of new conflicts.

Following the "Green revolution" during the 1970s, India achieved food self-sufficiency. In recent years, it has started to import food grains. Per capita arable land in India declined from 0.2 hectares in 1981 to 0.15 hectares in 2005. The government has emphasized the need to increase land productivity to boost food production, but serious doubt exists whether this is achievable.[134] Increasing land salinity is one of the critical elements, and there are also other concerns. Amongst many, a recent Finish study concluded that "the present climate models underestimate the increase of carbon dioxide emissions from soil in warmer climate. Thereby they also underestimate the accelerating impact of the largest carbon storage in forests on the climate change. This result is also essential with respect to the climate policy measures concerning forests."[135]

Despite the severe decline in land quality during the last two hundred years, Indian policymakers have not addressed this issue with the seriousness it deserves. Disease, environmental damage, deforestation, and bad agricultural practices have aggravated the situation,[136] affecting the ecological balance, particularly in the densely populated Gangetic delta region and in Bangladesh.[137]

The root cause of land salinity and degradation in India was the introduction of canal irrigation by the colonial administration, despite warnings from British engineers and soil scientists.[138] The situation has worsened in large parts of Punjab and Haryana where a short-term view of production has led to excessive use of fertilizer and salination of cropland. More than ten million food-producing acres have been lost to salinization due to improper irrigation.[139] Experience from Pakistan and parts of northern India points to an irreversible salinization of land due to increasing canal irrigation. Efforts to reclaim soil are being made, but with little success[140] and at high cost.[141]

The colonial authorities ignored the advice from local farmers in the name of scientific approach.[142] Without this local knowledge, the administration failed to notice the gradual deterioration in soil capacity, increased salinity, and related problems. Satellite imagery now reveals what is apparent at ground level. The soil in the Ganges plains in Haryana and Uttar Pradesh is deteriorating.[143] India's national water policymakers have not shifted from this British tradition of ignoring grassroots people who have immense practical knowledge.[144] It is difficult to say who they were referring to, but the global business leaders at the World Economic Forum in Davos in early 2010 agreed that water only entered the policy stage by the back door, and that water was politics, a claim that could well be directed at India when one considers its so-called national water policy. The forum's attendees virtually agreed that the rapidly deteriorating water environment could lead to economic collapse.[145]

Several fault lines crisscross India's internal sociopolitical scene. It has deep divisions in which factions had only been united by their opposition to foreign rule during the nineteenth and twentieth centuries.[146] At the end of colonial rule, the euphoria of independence and a dream of a new India carried the country, but cracks began to appear. These fault lines are widening and the water crisis will test India's underlying strength. The armed Maoist movement in parts of rural India has implications for national water policy. Indian government declared the Maoist a terrorist grouping in June 2009. Experience tells us that unless the root causes of poverty, hunger and injustice are addressed, such movements will continue to gain strength.

Internal political divisions inhibit a realistic freshwater policy, and resolution of any interstate conflict. The outcome of India's 2009 federal election indicates a stable federal government, but it does not necessarily guarantee an easier path to resolving interstate water-sharing disputes. Coalitions rule many of India's states, and often the dictates of political survival encourage political parties to opportunism.

The decade-long water-sharing dispute between Karnataka and Tamil Nadu, and conflicts between Punjab, Haryana, Rajasthan, and Uttar Pradesh, or the conflict between Maharashtra and Andhra Pradesh are examples. The list of conflicts is too long. States do not hesitate to accuse others of illegally harvesting water from common sources. The aggrieved states pressure central government politically for redress, but even this step has limitations.[147]

Most Indian coalition partners are motivated by the lure of the fruits of incumbency, and rarely by any philosophical underpinning or long-term national policy perspective. For example, one of the major coalition partners (Trinamul Congress, primarily a regional party but the government's survival depends on its support) in the national government is vehemently opposed to water pricing. By refusing to price water, the political leadership is assisting those who can afford to pay, and not helping those who are impoverished and have no access to freshwater. Independent estimates indicate that somewhere between two thousand and three thousand farmers committed suicide in the state of Andhra Pradesh over the last two decades.[148] In such circumstances the state rights issue becomes an effective foil. Political leaders often solve domestic problems by creating outside enemies.[149] Mahatma Gandhi's political brilliance was evident in selecting the salt issue as a weapon against the British in 1930 and confirms the importance of addressing issues in their wider context.

Issues of concern are many; in this chapter we have highlighted only few important ones, which need to be addressed immediately to bring under control the food and security uncertainties.

NOTES

1. L'vovitch,1979; Vőrősmarty et al. 2000.
2. Strong and Goransson 1991.
3. Black and King 2009: 25.
4. World Bank 2004: 5.
5. Ingram, Whiteley, and Perry 2008: 12.
6. Wolf et al. 2009.
7. United Nations 1995: 80–81.
8. Seregeldin 1995; Guest 1995; and Nilekani 2007.
9. Iyer 2000: vii; Iyer 2002: 299; Wolf et al. 1999.
10. Brichieri-Colombi 2009: 291.
11. Clarke and King 2006: 11; Peterson 2007.
12. Ingram (1990: 2) made this observation following her detailed examination of the political intricacies that were involved in the formulation of the Colorado River basin legislation in the United States.
13. Billon 2005: 8.
14. Gleick 2003; Strauss 2002:11.
15. Rothfeder 2001: 13.
16. Plaut 2000: 19–20; Gleick 1992.
17. Lowi 1999.
18. Gleick 2006: 192–218.
19. Wolf, Kramer, Carius, and Dabelco 2009.
20. Wade 1988: 221.
21. *Far Eastern Economic Review* 1988; Phadnis 1990: 255; and Sobhan 2000.
22. An investigative report by PBS *NewsHour* senior journalist on April 2, 2010.
23. Pomeranz 2009.
24. Conflicting reports are appearing in the Indian media. For example, the Indian English dailies *The Hindu* and *The Statesman* have both published reports concerning this issue. *The Hindu Weekly* (November 15, 2003) published that the Indian Ministry of Water Resources has requested the Ministry of External Affairs "to seek details" about recent news reports that China planned to divert waters of rivers originating in Tibet, including the Brahmaputra (known as Yaluzangbu in China). The *Statesman* also reported on the same day that Ms. Zhang Qiyue, Chinese Foreign Ministry spokesperson remained vague on the issue of building a dam over the Brahmaputra River in Tibet. She is reported to have said that China does not have any plans to build a power plant on the river, while *The Hindu* report claimed that a 40,000 MW plan was being considered. It is worth noting that China's approach to information dissemination is not as open as in countries with a democratic setup. In the past it was its practice to disseminate news for wider consumption well after the events had taken place.
25. The recent revival of the border disputes with India is another example. "Nepal Revives Border Feud with India, *Times of India*, March 18, 2010.
26. "Nepal Scraps Passport Deal with India to Maoist Pressure," *Times of India*, April 13, 2010.

27. Robert 1994: 30.

28. Gleick 1990: 50.

29. China's position with respect to becoming a member of the reconstituted MRC in 1995 provides an example: Although the country owns about 48 percent of the Mekong basin, it has remained an observer in the commission, making it much easy for it to use the Mekong waters for its own benefit, sometimes at the cost of the lower riparian countries.

30. Watagawa 1995: 32.

31. The Nile Basin countries have been making noises for renegotiating the Nile water-sharing agreement of 1929 for decades that would reflect the new realities. Egypt has been insisting on maintaining its 55.5 billion cubic meters of water from the river, more than half of the Nile's flow. It also wants veto power over any new irrigation projects undertaken by the other basin states. Following a meeting of member countries in Egypt, the sub-Saharan states have rejected Egypt's position and called for new allocations of Nile water to reflect their burgeoning populations and industrial capacity. Except Egypt and Sudan all other member states decided to sign the agreement by mid-May 2010. It was unclear what effect, if any, the signing would have on actual water exploitation (Nasrawi 2010). Also see (Nkrumah 2004).

32. Ward 1997: 67–68.

33. Briscoe and Malik 2006: xix.

34. Ray 2008a: 127–40.

35. "The Politics of Water," World Economic Forum 2009, http://www.weforum.org/en/knowledge/Events/2009/AnnualMeeting/KN_SESS_SUMM (accessed February 23, 2010).

36. Briscoe and Malik 2006: xix.

37. Chapman and Thompson 1995: 187.

38. Black and King 2009: 28; Clarke and King 2006: 81.

39. Qureshi 1981.

40. Ali 1942: 3; Ali 1949.

41. Klare 2010. Also, see Klare 2001.

42. Libiszewski 1994: 196–97.

43. Ohlsson 1995: 27.

44. Barlow and Clarke 2002: 72.

45. Hundley 1966: 180–85.

46. Libiszewski 1999: 123–28.

47. Rosegrant, Cai, and Cline 2002: 206.

48. China is facing a severe water crisis in and around Beijing, the industrial city of Tianjin, and other parts of the North China Plains. Some one hundred Chinese cities and towns, mostly in the northern and coastal regions, have suffered severe water shortages in recent years. It is estimated that in Beijing Central Urban District alone, water demand will increase by about 38 percent, surrounding rural districts by 12 percent, and Tianjin by 36 percent before the end of this decade (Science and Technology Commission, Peoples' Republic of China 1991).

49. Diamond 2005: 359, 364.

50. Friedkin (1987: 393–414) was a high-ranking American official. During his tenure he was involved in US-Mexico water dispute negotiations.

51. Austin 2004: 272.

52. Tanzi and Arcari 2001: 302.

53. Steiner 2010.

54. Stephenson 2009.

55. Assuming 2008 dollars. Costanza et al. 1997.

56. Libiszewski 1999: 117.

57. http://bicn.com/acic/ (accessed November 29, 2004).

58. http://www.censusindia.net/t_00_001.html (accessed June 28, 2004). The number of affected districts should be taken as indicative only. Government of India data show "place of occurrences." The places are identified by district names only. The number of affected districts is based on this assumption.

59. Freeze 1997: 149.

60. McAllister, Hamilton, and Harvey 1997: 1–140.

61. Answers are required on number of issues, such as role of microbes in the control of chemical reactions beneath the earth's surface, in both unsaturated and saturated zones; transport of chemical constituents through the unsaturated zone; hydrology of fractures rocks, flow, and transport (Bredehoeft 1997: 60).

62. Barlow and Clarke 2002: 30.

63. Clarke and King 2006: 41.

64. World Bank 2003: 3.6.

65. Kathuria and Gundimeda 2002: 141.

66. "Where Did Ganga Funds Go?" *Statesman*, February 21, 2007—a bench of the Supreme Court wants to know.

67. The Supreme Court passed this judgment on October 13, 2009. It imposed heavy fines on the garment factories in Tirupur, which were discharging "sludge" in the nearby Nayyal River.

68. "Pakistan's Water Crisis," *PRI's the World*, April 13, 2009. ("Global Perspectives for an American Audience": *PRI's the World* is a one-hour weekly radio news magazine. It is a coproduction of WGBH/Boston, PRI, and the BBC World Service, airing weekdays on over two hundred stations across the country). Cited in Pomeranz 2009.

69. Clarke and King 2006: 65.

70. Rosegrant, Cai, and Cline 2002: 6–7.

71. http://www.earth-policy.org/Indicators/indicator7.htm. (accessed February 10, 2004).

72. http://www.irc.nl/page.php/405. (accessed February 10, 2004).

73. Rosegrant, Cai, and Cline 2002: 3–4.

74. Barlow and Clarke 2002: 18.

75. Conniff 2007: 28–30.

76. Ingram, Fieldman, and Whiteley, 2008: 237; Ingram, Whiteley, and Perry 2008.

77. Pachauri 2008.

78. "Dangerous Weather, Climate Change Is a Security Issue the World Must Confront," *Financial Times*, April 19, 2007.

79. United Nations 2008.

80. United Nations World Water Development Report 2003: 12.

81. Worship 2009.

82. Shearman and Smith 2007: 43.

83. Schindler 2003: 164–68.

84. Mendelsohn, Dinar, and Sanghi 2001: 85–101.

85. Suri 2008.

86. Mendelsohn, Dinar, and Williams 2006: 159–78.

87. Hansen 2009.

88. *Economist*, March 20, 2010.

89. The climate change congress was held in Copenhagen and attended by more than two thousand scientists and environmental activists.

90. In a recent study carried out by scientists at the University of California, Davis, confirmed that

> climate "Tipping-Points" may arrive without warning. The scientists further claimed that it is harder than experts thought to predict when sudden shifts in Earth's natural systems will occur—a worrisome finding for scientists trying to identify the "Tipping-Points" that could push change into an irreparable global disaster. (Hastings and Wysham 2010)

91. Musil 2009: 21.

92. Press reports indicate that capturing carbon dioxide from the atmosphere and storing it in deep geologic formations is also a viable option, and few rich countries have already claimed to have this mechanism operating. The Japanese project Mikawa power station near Fukoka Prefecture, Sleipner field in the North Sea where the Norwegian group Statoily and one project each in Canada and Algeria is also reported to have started using this technology. These claims need further investigation, particularly with respect to cost, major safety issues, and complex sociopolitical issues. The United States is reported to have invested $2.4 billion in the technology and hopes to have the technology of carbon capture ready for its use by 2016, according to the energy secretary Steven Chu (Dyer 2008: 203).

93. "Block the Sun, Control Global Warming," *Times of India*, February 1, 2010. Canadian and US scientists want to block the sun to cool the earth and limit global warming.

94. "Every Silver Lining Has a Cloud," *Economist*, January 31, 2009.

95. Kelly 2003; Olson 2004.

96. Pelåez and Pelåez 2008: 123–24; Casey 2007.

97. Professor Scott Barrett, director, International Policy Program, and Johns Hopkins University in Washington, DC, at a group discussion, "The Climate Engineers," April 6, 2008, www.abc.net.au/rn/backgroundbriefing/stories/2009/2444935.htm.

98. Professor Chris Field told at an American science conference in Chicago. The report was posted by the Australian Broadcasting Commission on February

15, 2009 (Borenstein 2007; Adam 2007); http://www.abc.net.au/news/stories/2009/02/15/2491942.htm.

99. The information was provided by the World Meteorological Organization and the International Council for Science on February 26, 2009; http://www.abc.net.au/news/stories/2009/02/26/2502076.htm.

100. *Economist*, December 20, 2008.

101. Two of their comments are worth noting: "This energy imbalance is the smoking gun we have been looking for" and "This is not a smoking gun; climate change is a battalion of intergalactic smoking missiles" (Musil 2009: 19).

102. Moy et al. 2009.

103. Greenwald and Kahn 2009: 167; Chanda 2009.

104. Greenpeace estimates that 80 percent of all life on Earth is found in our oceans. Oceans provide over half of our life-giving oxygen and vital sources of protein for humans, drive weather systems, and supply natural flows of energy and nutrients around the world (publicity material issued on February 22, 2009).

105. http://www.undp.org/cpr.

106. A "tipping-point" is defined as the point where a small increase in temperature or other change in the climate could trigger a disproportionately larger change in the future. The study team of thirty-six leading climate scientists was headed by Professor Timothy Lenton of the University of East Anglia.

107. The Indian monsoon relies on temperature difference between land and sea, which could be tipped off balance by pollutants that cause localized cooling. The other tipping-points are the Arctic, the Greenland and the West Antarctic ice sheet, the Gulf Stream, El Niño, the West African monsoon, the Amazon rain forest, and the Boreal forests (Connor 2008).

108. Clarke and King (2006: 11), For a diagrammatic exposition of the tipping-points, see Gleick (1990: 41–43).

109. Vickers 2005: 178–84; Polsky and Cash 2005: 215–36; and Wilhite and Pulwarty 2005: 395–96.

110. See "Public Health Agency, Canada," cited in Palaniappan, Lee, and Samulon 2006.

111. Barlow and Clarke 2002: 208–9.

112. Rashid, 2000: 183–95 and 207–16.

113. Rahman 1999: 208.

114. Doug Benn, "In Himalayas, Evidence of 'Alarming' Glacial Melting," in an interview with James Mates of Independent Television News, aired by *NewsHour* on December 15, 2009; http://www.pbs.org/newshour/bb/environment/july-dec09/everest_12-15.html.

115. "Pachauri Admits Mistake In IPCC Report, Rules Out Resignation" and "UN Climate Panel Head Says Glacier Alarm Regrettable Error," *Times of India*, January 25, 2010.

116. "Melting Himalayan Glaciers Threaten 1.3 Billion Asians," *Times of India*, January 26, 2010 .

117. "Himalayan Glaciers Have Shrunk by 38 Percent in 40 Years," *Times of India*, January 26, 2010.

118. "Govt Findings on Himalayan Glaciers Misleading: NGO," *Times of India*, January 26, 2010.

119. "Gangotri Glacier Dying a Slow Death," *Times of India*, January 19, 2010.

120. Aditi Kapoor, "Scientific Behaviour on the Thin Ice," *Times of India*, January 20, 2010.

121. Pomeranz 2009.

122. "Himalayan Meltdown Catastrophic for India," *Times of India*, April 3, 2007; "World's Top 10 Rivers at Risk: WWF Report," *Hindu*, March 21, 2007; "The Great Himalayan Meltdown," *Hindu*, April 10, 2007; C. Buckley, "Global Warming Imperils Himalayan Glaciers," Reuters, April 23, 2007; "Warming a Threat to Natural Wonder," *Times of India*, April 6, 2007; M. Bristow, "Melting Himalayan Glaciers Threaten Chinese Water Supplies," *Kyodo News*, March 22, 2007; and Richard Spencer, "Shepherds Delight as World Worries about the Big Melt," October 6, 2004, http://www.abc.net.au/news/newsitems/200410/s1213705.htm.

123. Huggler 2007; United Nations 2006: 166; Dhakal 2003.

124. Mukherjee 2001; Wade 1985: 485.

125. "The Shrinking Alps" (Australian Special Broadcasting Service report), http://www.sbs.com.au/dateline/story/transcript/id130774/n/The-Shrinking-Alps (accessed December 21, 2009).

126. See Hansen 2006b.

127. Dasgupta 2007.

128. James 2009: 68.

129. Pel ez and Pel ez 2008: 123.

130. Misra 1970: vii.

131. Sen Sharma 1997: 175–82.

132. World Resources 1992: 1; Postel 1992: 28–37.

133. National Family Health Survey of 2005–2006.

134. "End 'Greed Revolution', Says M. S. Swaminathan," http://www.thestatesman.net/page.news.php?clid=2&theme=&usrsess=1&id=47855 (accessed July 18, 2004).

135. Karhu et al. 2010.

136. Goldsmith and Hildyard 1984: 27–48; Barber and Ryder 1993: 21.

137. Crow 1995: 129–36.

138. Whitcombe 1972: 72; Pearce 1992: 67–71.

139. Simon 1998: 68.

140. Burke 2002.

141. British Broadcasting Corporation, "Feeding the World," 2007; the BBC World Service series investigated the growing but often-underreported challenges facing world's food supply. The reporter went to Punjab's rural heartland and found that many farmers were in a desperate situation because of rapidly declining water levels and land salinity ("Growing Pains, Part I") (Dubash 2002: 150–51, 249; Barbier 2005: 288); http://news.bbc.co.uk/2/hi/programmes/documentary_archive/6500041.stm (accessed March 29, 2007).

142. Gilmartin 1994: 1137.

143. Chapman 2000: 109; Chapman and Baker 1992: 30–31, 61–62.

144. Sengupta 1985: 1919–38; Mitra 1996.
145. "The Politics of Water," World Economic Forum 2009.
146. von Tunzelmann 2007: 16.
147. *Statesman*, April 12, 2007.
148. Bellman 2004; Roy 2004. "Andhra to Probe Farmers' Suicide," *Statesman*, July 13, 2004.
149. Simon 1998: 65.

Chapter Three

Cold War: Mutual Suspicion and Political Psyche

The post-1947 era remains one of the most eventful in the subcontinent's history. India became independent and Pakistan came into existence. Lord Curzon partitioned Bengal in 1905, but it was reunited by King George V in 1911 following fierce protest against the partition.[1] Curzon's division irreversibly damaged community thinking and the colonial power encouraged the growth of a feeling of separate identity.[2] The British establishment steadfastly opposed India's independence, but eventually granted a fractured split of India and created Pakistan.

The Rana dynasty in Nepal lost power in 1950, Bangladesh was created in 1971, and the Chinese Revolution led to the birth of the Peoples' Republic of China in 1949. Later, the decline of the Cultural Revolution in the 1970s saw the ascendency of modern China. Most importantly, the Cold War reached its zenith during the 1960s and profoundly influenced intercountry relations in the subcontinent. The Maoist-led arms struggle began in Nepal, and eventually led to the abolition of the monarchy and the establishment of Nepal as a republic on May 28, 2008.

Following the end of the Second World War in 1945 and the start of the Cold War, ideological rivalry between the United States and the former Soviet Union influenced the policy directions of most countries, including India and Pakistan on almost all political, economic and security issues. These two countries fought three wars during this period, but with no decisive winner (although the 1971 war led to the creation of Bangladesh—previously a part of Pakistan). Relations between India and Pakistan have never been cordial and remain fractured to this day.

Pakistan was created as a separate homeland for the Muslim population, but a significant proportion of Muslims opted to remain in India. Indeed, when the Punjab was to be partitioned in 1947, 90 percent of its Muslims voted against dividing the province.[3] Since its birth, Pakistan has been ruled for longer periods by the military than by civilian rulers. Pakistan's security concern revolved around its survival, as it believed its very independence would be threatened by the Indian hegemon. This position indirectly contributed to Pakistan's unquestionable alliance with the United States during the Cold War. In the end, religious bonds alone could not keep the country united, and the war of 1971 led to the establishment of Bangladesh.

During the Cold War, India's security objective was to prevent foreign powers intruding into Indian affairs, and to limit their influence in India's immediate neighborhood. This thinking tilted India toward the Soviet camp, and the Soviet Union steadfastly maintained its political and economic support for India's needs. Although India remained nominally with the Non-aligned Movement, its economic and defense closeness to the Soviet Union contributed to worsening security concerns, as the United States and its allies took every opportunity to ensure that Pakistan remained an ally, as an insurance against the spread of Soviet influence. Consequently, two security agreements sponsored by the Western Block—the South East Asia Treaty Organisation (SEATO) and the Central Treaty Organization (CENTO) found easy acceptance in Pakistan.

SEATO and CENTO were respectively sponsored by the United States and the United Kingdom. The objective of SEATO was to protect Southeast Asia from the spread of communism. SEATO ceased to exist after the end of the Vietnam War in 1975. CENTO was never a success, as many viewed it as an attempt by the British to retain influence in the Middle East as a substitute for the loss of their Indian empire. These contributed partially to deteriorating relations between India and Pakistan.

The location of Pakistan and the generally obsequious and compliant nature of the Pakistani military and diplomatic elite provided an excellent opportunity for the West to achieve its political objectives during the Cold War. This kept Pakistan on tenterhooks and denied it any opportunity to nurture its civil society institutions. Consequently, Pakistan still struggles to find a firm footing, while India in 1952 became the world's largest democracy at a time when the United States still had in place its Jim Crow laws against black citizens.[4] The long-term legacy was bad blood between two neighbors; subsequent events only served to reinforce this.

Following the China-India border conflict in 1962, Pakistan found in China an important new ally; and the Chinese nurtured the relationship to the fullest. This close relationship was evident when China used the American embargo on arms sales to Pakistan following the 1965 Pakistan war with India to reverse American policy regarding China.[5] Although this seems at first to be a paradox, as with most countries China's national interests take precedence over other concerns. The strategic importance of this stance is considerable in any future dispute or conflict between India and Pakistan, be it on issues such as building new dams on the Indus River or on its tributaries or the future of Kashmir.

The Indus River water-sharing issue became contentious shortly after independence, but common sense prevailed and the countries agreed to accept the World Bank–sponsored Indus Waters Treaty of 1960 as a basis for settling the water-sharing arrangements. The World Bank provided funding so that the parties would find minimal or no difficulty in executing the provisions of the Treaty.

Water-sharing arrangements in the eastern part of the country did not become a critical issue immediately after independence. The Pakistan leadership was more concerned with the well-being of the western part of the country. The water-sharing issue only became contentious when India started construction of the Farakka Barrage in 1960. Finished in 1974, the dam in West Bengal is close to India's border with Bangladesh. Notwithstanding the formal agreement reached between India and Bangladesh on sharing the water from the Ganges River, the water-sharing conflict continues to surface regularly due to lack of water, particularly during the summer months. Access to water in the east and northeast therefore remains of major concern and a security issue for both countries. The end of the Cold War did not diminish water's importance as a security issue.[6] As the world is now preoccupied with environmental issues influencing water crises, water remains as critical an issue as before.

As the water crisis in the subcontinent deepens, past legacies cast shadows on the policy directions of the countries in the region. The issues are complicated because of the ethnic, social, political, and economic complexities within each country, including the complex nature of the insurgency movements in India, Pakistan, and Nepal.[7] Insurgency movements have their own dynamics and water-related conflicts are often a boon for them, as such movements recognize that unfettered access to water is a cause close to the heart of common people.

This is important from a policy perspective, as experts broadly agree that water resources are managed in a wider socioeconomic context, and not exclusively on the geographic limits of the river basin.[8] Sociopolitical-socioeconomic dynamics and sovereignty issues usually determine a country's foreign policy directions, and these factors have a strong bearing on transboundary water-sharing arrangements. This is happening in the subcontinent where countries have been squabbling and fighting in total disregard of their history.

The long decolonization process has left a permanent scar in the minds of the citizens of these countries. The fear that India wanted to dismember Pakistan (particularly after the creation of Bangladesh) and then reabsorb the parts into the Indian state has dominated Pakistan's policy thinking toward India, although other internal and external factors (such as the Kashmir conflict) have contributed to this suspicion and fear. At various stages, other neighbors have had similar concerns about India's power in the region.[9]

Although Bangladesh came into existence with India's active military support, good relations never flourished. On the contrary, water-sharing conflicts, infiltration of illegal migrants to India, and providing shelter to Indian insurgents in Bangladesh have contributed to ongoing poor relations between these two countries.

Anti-Indian feeling in Nepal has its roots in history, and partially in its geographic location: Nepal's dependence on India as an exit route to the outside world, its mass poverty, and India's overt influence on Nepal's governance policies have all contributed to an anti-Indian feeling, particularly among the middle class. The 1990 Act concerning Nepal's natural resources (including water) is a manifestation of this feeling.

Collectively, the enormous difference in physical, demographic, and economic sizes between India and its immediate neighbors has adversely contributed to these complex relationships. Mutual distrust is also a product of the policies of colonial rulers that continued into the postcolonial period. Suspicion compounds the existing complexities and uncertainties that respective governments face in managing complex interconnected systems and issues relating to water-sharing arrangements.[10] Due to mental blockages and mutual distrust, even small events have often assumed disproportionate importance in resolving problems.

In a rapidly changing globalized world, where mutual dependency provides opportunities for best policy outcomes, the countries of South Asia are performing poorly. Many of their disputes have little relevance in Europe where there exists mutual recognition of each European nation's role in the regional geopolitical-geoeconomic environment. Europe's mutual understanding has been achieved over years, as distinctive changes have taken place in the social conditions of humanity, beginning with the French Revolution,[11] and then with the momentum created by the decolonization process. The end of the Cold War further contributed to the growth of this mutual understanding. Political egalitarianism in Europe has also made it possible for smaller European countries to develop their independent positions in the post Cold War period, based on democratic pluralist societies.

Notwithstanding India's adherence to principles of democratic governance, political egalitarianism is only partial in India. In Pakistan and Bangladesh, it has never existed. The situation is further complicated by religious elements that have dominated the political thinking of leadership, the absence of functioning democratic institutions, and the existence of three sets of asymmetrical relations (discrimination, exploitation, and oppression).

In the 1990s the Nepal's elected government, under the country's new constitution, was plagued by infighting among coalition partners. Now Nepal faces an uncertain political environment; first, with Parliament's formal abolition of the monarchy in December 2007; and, second, by the political uncertainty created by the Maoist political objectives, which are often incompatible with the political objectives of other participating political parties in the ruling coalition. Despite not participating in the current coalition government, the Maoists are the largest single party in the Parliament.

Bhutan has a newly elected government under a recent constitution that formally established a constitutional monarchy in 2008. It will be some time before one can pass judgment on this new development.

Situations in Bangladesh and Pakistan are complex. Their internal socio-political-ethnic complexities have clouded each government's approach to contentious issues such as water, as they position themselves to extract maximum political advantage from any conflict, often irrespective of merit or justification. The history of the subcontinent since independence does not offer grounds for optimism. Collectively, all nations in the region have contributed to the growth of distrust between big India and its relatively small neighbors.

The biggest impediment to developing a mutually acceptable and sustainable water policy is the suspicion that exists between the riparian countries, at the political and community level. The most important strategic concern for all these countries is how to overcome this distrust so as to resolve existing and emerging conflicts and disputes.

THE RISE OF MODERN CHINA

The rise of China as an economic and political force has created both complexities and opportunities, particularly for the smaller countries in the region. Some commentators believe that high rates of economic growth with increasing exposure to global political realities and international market forces will force both India and China to cooperate and settle unresolved disputes. If past experience is any guide, such possibility appears to be somewhat remote, as one cannot ignore the fact that two big powers have never existed side by side in Asia and flourished at the same time.

Fundamental differences exist between China and India as regards their respective governance systems and international policy stance. Moreover, there are two distinct differences between these neighboring giants: first, China controls the water sources of major Himalayan rivers that flow through the subcontinent; and, second, China has shown a unique capacity to carry small states with it when its national interest is at stake.

India and China are yet to resolve their fifty-year-old border dispute, although China has resolved matters with Nepal. Notwithstanding verbal commitments by both countries to settle their border disputes peacefully, China has created new areas of dispute by claiming that one of India's northeastern provinces belongs to China. In addition, from time to time China has continued to send out conflicting signals on its position with respect to the border issue. The same is true with the Brahmaputra River water-sharing issue, where there have been recent developments (this issue is discussed in chapter 5).

Both India and China are water-scarce countries and it is natural for them to harvest freshwater, where feasible and available. In this activity, China has a natural advantage and a political advantage. Two of the three constantly flowing rivers in India originate in China and provide that country with opportunities to harvest waters from them. Being a one-party state, China has the capacity to ignore all opposition to government policies. This factor makes it easy for the government to implement projects that face staunch opposition, whether that opposition is internal or outside of China. This is unlikely in India because of its democratic governance structure, in which checks and balances often make it difficult for governments to carry out their programs without community scrutiny, at least theoretically, both at the national and international levels.

China and India pursue entirely different political paths and, in practical terms, policy success has a somewhat different meaning in each country. The philosophical underpinning of their respective policies is different, and diverging political and economic ambitions, increasing rivalry over market shares, access to natural resources or selection of strategic partners could drive them further apart.[12]

The end of the Cold War and economic globalization has led to fundamental shifts in the conceptual thinking on strategic and security issues in most countries. It is generally agreed that the contemporary security agenda is founded in the dissolution of international politics that has attended the opening up and subsequent interdependence of national economies.[13]

Issues of political economy have become elements of the new security architecture, rather than remaining as problems subject to the calculus of welfare maximization. These issues, and the threats created by climate change, imply that water no longer remains an ancillary issue on the periphery of the development and policy debate. Rather, water has become a centerpiece policy issue in each country's strategic considerations.

NOTES

1. Puri 2009.
2. Chapman 2000: 3–26, 295–96.
3. Puri 2009.
4. Roy 2007.
5. Chapman and Baker 1992: 70.
6. Blatter, Ingram, and Levesque 2001: 38.
7. In his book *Climate Wars*, Gwynne Dyer (2008: 113) cites an example that clearly shows the extreme nature of the sensitivities of some people in the subcontinent on water issues. Dyer writes that at a 2001 seminar in Karachi, Pakistan, tensions only arose when someone alleged that the Indian government had plans to use the water "weapon." At this point, a participant warned that any conflict over water would lead to Pakistan using nuclear weapons on a first-strike basis against India.
8. Brichieri-Colombi 2009: 11. See also J. A. Allan 1999.
9. Bradnock 1992: 70.
10. Waldrop 1992: 66; and Nye and Donahue 2000: 11.
11. Gellner 1999: 179–224.
12. MacDonald 2007; Bakshi 2004: 274–75.
13. Bayne 1994: 7; Crawford 1994.

Chapter Four

Critical Policy Concerns

The short- and long-term impacts of environmental crises on global economy, including freshwater resources, are enormous. Water needs special consideration because, first, without it no life can exist and, second, although almost every other resource has an alternative, this is not the case for water. Because of water's all-inclusive nature; public policies, private and nongovernment sector activities, and those of the entire community, must ensure that each step passes stringent environment standards on water loss/gain. This requires consideration of climate change and water issues in tandem with other policies to ensure that all policies, as far as feasible, satisfy community, national, regional, and international obligations.

Some climate change consequences can be predicted: some effects are hard to foresee. For the predictable consequences, governments and communities will have time to prepare responses and develop strategies. In many instances negative consequences of climate change damage cannot be predicted sufficiently in advance, but may require rapid response to avoid disruption and suffering. Further, all policy responses to such situations need to be consistent with national conditions and community policy priorities as far as practicable, although, for some cases, policy action at regional or international levels will be a priority. In a climate change context this issue becomes particularly important, as climate change does not remain confined within political boundaries.

The policy concerns of any country today are multifaceted, such as environmental degradation, climate change, and the finite sources of energy, globalization, economic interdependency, and terrorism. Any lingering doubt about the nature of interdependencies carries no weight, as is clear from the global financial crisis of 2008–2009. Although the crisis began in the United States, other nations are not immune from the consequences.[1] It had a domino effect, with one commentator saying in jest, "The world's 1 billion meat eaters and car drivers are responsible for the lion's share of the damage humans have caused to common global resources."[2]

Generally speaking, climate and water policies impact on every aspect of our daily life. The world needs to address many other issues, but this book focuses on four issues that have enormous consequences for everyone on the planet, irrespective of their economic or social status. Leaving out energy, the identified critical issues are environmental displacements, disease, food crises and internal and external security. Climate change affects countries and individuals in different ways, depending upon their individual perceptions and situations.[3] A global survey[4] in 2007 found that the biggest concern of 48 percent of the global population was that the next generation would live in a less safe world.[5] As such this has no practical bearing, except that the term "safe world" means different things to different people, though a common thread of survival binds most of the concerns.

The survey was conducted by the World Economic Forum and its objective was to establish, for politicians and business leaders, peoples' perceptions of priority issues. Out of thirteen identified issues,[6] five priority issues for India were the war on terrorism (19 percent), protecting the environment (16 percent), closing the gap between rich and poor countries (15 percent), encouraging economic growth and improving the global economy (11 percent), and eliminating extreme poverty and hunger in the world (10 percent). For Pakistan, priorities were eliminating poverty and hunger in the world (24 percent), the war on terrorism (14 percent), closing the gap between rich and poor countries (13 percent), encouraging economic growth and improving the global economy (12 percent), and reducing wars and conflicts (10 percent). Detailed data for Bangladesh, Nepal, or Bhutan was unavailable. There are two major differences between India and Pakistan with respect to two of the issues. First, protecting the environment was a top priority in India, but hardly received a ranking in Pakistan (only 3 percent) and second, only 10 percent of Indian respondents identified extreme poverty and hunger as the principal issue, while nearly one-quarter of the Pakistan's population focused on this issue. This particular finding is consistent with those looking at Pakistan watchers who believe Pakistan's economy does not serve its population. Their view is that a small group of elites controls the economy to serve their own interests.[7]

Bangladesh, India, and Pakistan have failed to provide most of their citizens with basic infrastructure on a sustainable basis. This could be interpreted as denying them their basic human rights.[8] Such failure has earned them the dubious distinction of being the collective depository of the largest number of people living below the poverty line. Globally about one person in five lives in poverty.[9] Collectively, the four subcontinental countries had about 451 million poor living on less than US$1 a day during 1990–2003. India alone had 371.6 million (34.7 percent of the population), followed by Bangladesh 49.2 million (36 percent), Pakistan 20.34 million (13.4 percent), and Nepal 9.8 million (37.7 percent). About 79 percent of the total population of these countries lived on an income of less than US$2 a day during the same period.[10]

Poverty in India became worse as a consequence of the global financial crisis at the beginning of the second decade of the twenty-first century. Contrary to popular belief that the crisis left India unscathed, the United Nations Department of Economic and Social Affairs (UNDESA) reports India as being the biggest victim of financial crises-induced poverty. It estimates that the number of India's poor was 33.6 million higher in 2009 than would have been the case if the growth rates of the year from 2004 to 2007 had been maintained. In 2009 alone, an estimated 11.6 million more people in India became poor or remained in poverty than would have been the case at 2008 growth rates.[11]

Climate change–related issues are neither simple, nor easy to address and, generally speaking, have economic consequences in terms of costs and benefits. If these issues are not addressed in a timely fashion and in a balanced way, the probability of internal and regional instability and turmoil will be high. The global information revolution has provided people with tools to assess what has been happening elsewhere and how people and governments in other places are addressing issues faced by their respective populations. In other words, we are now living in a globalized world and cannot remain outside this sphere.

Globalization has its various facets including economic, political, and cultural. During the Bengal famine of 1943 some four million people died, although there was no overall shortage of rice in British Administered Bengal.[12] On being appraised of the situation, Winston Churchill is reported to have asked if food was so scarce, why had Mahatma Gandhi not died yet?[13] If this were to happen now, Churchill's fellow citizens would be the first to demand assistance be sent immediately to mitigate the suffering and distress of the affected population. In 1943, this simply did not happen.

There has been a fundamental shift in global thinking in the early part of the twenty-first century. Shared global concerns for issues such as human rights, environment, and governance practices have revolutionized the perception of global citizens, inspiring them to organize and mobilize public opinion outside the formal structures of government. Even the most recalcitrant of governments often find it difficult to ignore such pressures.

ENVIRONMENTAL REFUGEES

Any climate change–induced rise in sea levels displaces people from low-lying areas.[14] The Intergovernmental Panel on Climate Change (IPCC) concluded in 2007 that expanding ocean water driven by climate change would drive up sea levels, on average anywhere from 18- to 59-cm by 2100, depending on the level of greenhouse gas emissions in the atmosphere. As sea levels vary across regions up to 24 cm, influenced by a process known as thermohaline circulation,[15] scientists now find that even Wall Street could be barely a meter above sea level as the century unfolds.[16] Climate change also contributes to increasing numbers of typhoons, erosion in coastal regions, and salination of inland freshwater and aquifers.[17] These all contribute to displacing people in large numbers from coastal or low-lying regions[18] and have already led to an increase in the number of involuntarily displaced persons (in this case environmental refugees).

The situation may worsen, following the recent ice bridge ruptures in the Antarctic. This breakup will have no direct impact on sea levels because the ice is floating, and does not add to sea levels when it melts. Like the Arctic ice cap, these ice flows displace their own volume. However, its impact on two fronts is more concerning. First, when ice shelves are not there, the glaciers and landed ice behind them start to move toward the ocean more rapidly. The UK-based Scientific Committee on Antarctic Research, a committee of the International Council of Science, predicts that about a third of all Antarctic sea ice is likely to melt, removing the buffers restraining giant glaciers from sliding into the ocean, which in turn could mean sea level rises up to six meters by the end of the century.[19] This would contribute to rising sea levels, although the actual level remains a matter of ongoing scientific debate.

On this count, two of the worst affected countries in South Asia are Bangladesh and the Maldives. Commentators argue that Bangladesh may be the place in the world most threatened by climate change.[20] Outmigration from the low-lying areas of southern parts of Bangladesh has already begun, albeit on a relatively low scale. Bangladesh has no land capacity to rehabilitate such large numbers of displaced persons. The Maldives has already lost a few of its small atolls under the rising sea level. Maldivian displaced persons cannot be rehabilitated internally, as there will be no Maldives if the entire country disappears under the waves.

Maldives president Mohamed Nasheed made a passionate appeal at the Copenhagen Summit for an appropriate global policy initiative, but without much success. In March 2010 he again made a desperate appeal at a climate change seminar in Helsinki and argued that his low-lying country is one of the most vulnerable to the rising sea levels anticipated as a result of global warming. Further, in a bid to lead by example, he pledged to make the Maldives carbon neutral by 2020 and addressed the international community: "The climate change debate should be reframed in economic and security terms ahead of a year-end UN summit in Mexico seeking a binding climate deal."[21]

Also vulnerable are parts of India that include the Indian Gangetic delta, and Mahanadi and Godavari deltaic regions, and low-lying coastal areas of Sri Lanka and Pakistan. The Gangetic delta has 104 islands with human settlements: India and Bangladesh share these islands on a 40:60 ratio. A few of these islands have disappeared over the past decade and another twelve remain vulnerable.[22] Parts of Ghoramara, a large Indian island in the delta region has already disappeared under water and some parts of it are lost to erosion.[23] It may vanish entirely in the near future.[24] Generally speaking, the low-lying areas of the deltaic region have not received the policy focus that they deserve in the climate change context. However, untiring efforts by some nongovernment organizations (NGOs) have compelled policymakers in recent months to direct some attention toward the plight of people in this region.[25]

British charity Tearfund predicts that there may be up to 200 million climate refugees in developing nations by midcentury.[26] A more alarming estimate comes from Jim Hansen.[27] He argues that if temperature increase follows prehistoric climate change patterns, the United States could lose most East Coast cities and even the entire state of Florida. In that nation, 50 million people live below sea level. The displaced numbers in China would reach 250 million, Bangladesh 120 million, and India 150 million.

Estimates of displaced persons by other commentators are lower. Alan Dupont and Graeme Pearman, and Al Gore separately estimate that the number of displaced persons in the Gangetic delta will vary between thirty million and sixty million.[28] The IPCC estimates that an additional to the unspecified number of current migrants, another thirty-five million will have crossed into India by 2050.[29]

Bangladesh has one of the highest population densities in the world. In 1999, its rural population density was 1,209 persons per square km compared to the world average of 524 (table 4.1). During 1980–2000, arable land per million of people declined substantially in Bangladesh and India, but in Pakistan the rate of decline was very high. In Bangladesh it declined from 103.98 hectares to 61.68 hectares, in India it declined from 144.8 to 97.20 hectares, and in Pakistan it declined from 241.5 to 153.5 hectares.[30] Large increase in population has contributed in this per capita decline in arable land. The governments in all these three countries need to critically look at their population policy along with their food policies to avoid future food shortage, because of

Table 4.1. Population density in mainland South Asian countries

Country	Population (2004) in millions	Density per km²	Arable land (1999) as a percent of land area	People per km² of arable land
Afghanistan	. . .	. . .	12.1	257
Bangladesh	140.5	1,079	62.2	1,209
India	1,079.7	363	54.4	444
Nepal	25.2	176	20.3	686
Pakistan	152.1	197	27.5	403
World	6,345.1	49	10.5	524

Source: Columns 2 and 3, *World Development Report 2006* (World Bank 2006: 292–93). Columns 4 and 5, *World Development Indicators 2002* (World Bank 2002: 134–36); for a definition of arable land, see page 137.

climate change and increasing population pressures on available arable land in their respective countries.

Bangladesh's high-population density problem is not new. Even when it was part of Pakistan (1947–1971), Pakistan campaigned for extra land from India to accommodate its growing population in the east.[31] One author even quotes from a document allegedly written by Seikh Mujibur Rahman, the founder of Bangladesh, making the same plea. This situation is the natural culmination of the way the Radcliffe Commission established the Radcliffe Line (the border between India and Pakistan in 1947).[32]

Bangladesh has a 4,000 km porous border with India, and illegal migration has been the pattern for a long time, although Bangladesh continues to deny it.[33] Even the United States with its 2000 km border with Mexico, finds it practically impossible to stem the flow of illegal migrants from the south, notwithstanding its resources to secure its borders against illegal immigrants.

Illegal migration across the India-Bangladesh border is already a politically explosive issue for India.[34] India claims the existence of a large number of Bangladeshi illegal migrants well beyond the border areas, in cities as far as Delhi and Mumbai; and links some of those migrants with recent terrorist attacks.[35] Although the Bangladesh leadership denies the existence of illegal migration, independent researchers and commentators tell a different story.[36] In reality, it is doubtful whether anybody knows the real number of illegal migrants in India.

There has also been significant Nepali migration to India since 1950 when the two countries signed the 1950 Indo-Nepal Treaty of Peace and Friendship permitting the free movement of people and goods between them. Nepali migration to India is not due to climate change. Nepal has a population density of 686 persons per km^2, yet only 17 percent of its land is cultivable. This situation and the endemic poverty have encouraged migration, which has caused political complexities in eastern India.

The involuntary displacement issue has assumed significant political dimensions in recent years for two additional reasons. First, it has become a human rights issue[37] strongly promoted, amongst others, by NGOs.[38] It impacts on the economy, society, and gender relations as a set of power relations operating at the level of the household, and its link to the outside world.[39] It also contributes to family disintegration.

Second, political complexities have arisen for two reasons. First, countries from whence large-scale environmental displacement takes place are among those least responsible for greenhouse gas emissions.[40] Second, but more importantly, those countries emitting a significant amount of greenhouse gases have indicated their reluctance to cut their emissions, saying they will not do so unless the developing country contributors also make the same level of reduction. The significant emitters are concerned about the effect on their domestic economies.[41] This was when the US Senate voted against the Kyoto Protocol in 1998, citing the exemptions of emission cuts for China, Mexico, India, Brazil, and South Korea as the major reason for its opposition.

Developing countries (particularly G77 countries) on the other hand (albeit with some support from few developed countries) argue that the average US citizen is responsible for emitting at least four times the level of greenhouse gases than their Chinese counterparts. The gap between the perception of industrialized countries such as the United States and other fast-developing countries such as China, India, Brazil, and South Africa is wide. Jan Pronk, a respected political commentator has even questioned whether the global community has approached the issue correctly.[42] He argues that the focus has been on procedures, but that emphasis needs to be on devising a core approach to mitigation, as other approaches will not do the job.

Migration from Bangladesh and Nepal will continue.[43] With long porous borders, weak economic structures and poor governance, people find it tempting to move to a place where there exists the possibility of a better life. As regards illegal Bangladeshi migrants, the cultural and linguistic similarities between Bangladesh, West Bengal, and parts of Assam are additional reasons to cross over into India. Would-be migrants also prefer places where they can expect to find cultural or clan affinity, as well as economic opportunity. In that environment, illegal migrants usually find little difficulty in assimilating within the local communities unless their presence is overwhelming, as is in Delhi or Mumbai. The large numbers of Bangladeshi migrants have created tensions in Assam, as it is alleged that the illegal migration there has been significant.

Usually such migrants come from the lower socioeconomic strata, and mostly get absorbed in the informal sector of the receiving economy. However, local communities with high levels of unemployed perceive immigrants as competitors creating sociopolitical tensions, as is the case in many European countries.[44] It is not possible for India to accommodate such large numbers of legal and illegal migrants. Besides India's high population density, its high unemployment rate remains a hidden time bomb.[45] Illegal migration has become a political issue, with some even saying that it impacts election outcomes.[46] The issue, therefore, needs to be considered at the international level.[47] Failure to initiate a collective policy action would only lead to complex socioeconomic and political problems, in the medium to long term.[48] Monetary investment alone is insufficient to address these issues.[49]

HEALTH

One of the least understood aspects of the impacts of climate change is its effect on human health. Clear links between climate change and the environment, or ecosystems upon which human health and well-being depend are yet to be fully understood by policymakers.

The impact of climate change on human health is enormous. Some impacts relate to extreme weather such as heat waves or cold snaps, while others relate to floods or droughts or other natural disasters such as wild fires and cyclones.[50] Yet another important effect that is often ignored or given scant attention is its impact on biodiversity. In the emotionally charged climate change debate this aspect received some attention, yet deserves to be studied further. Dr. Eric Chivian, a leading researcher on health and the global environment,[51] recently drew considerable attention when he warned, "Global warming presents a clear and present danger to biodiversity, the broad but ever narrowing range of plants and animals on Earth."

Chivian argued that a large number of species have evolved unique physiologies and biochemistries over millions of years that have an enormous amount to teach us about how our bodies work. He cited examples of tadpoles, polar bears, cone snails, and a tree called the Pacific yew, each of which has singularly made enormous contributions in understanding some of the most dreadful diseases faced by the human race thus far. Peptic ulcers, osteoporosis, type 2 diabetes, and ovarian cancer are just four examples.[52] In his keynote address at an international conference in Melbourne, Australia, Dr. Aaron Bernstein, another well-known expert, based at the Harvard Medical School made a similar point by stating that we need to pay serious attention to the implications of climate change on human health.[53]

The situation has already reached a critical level in India, although the Indian government is trying to relocate the endangered species from their usual habitat to protect them. The Botanical Survey of India, an institution set up by the Indian government in 1887 to survey the plant resources of the Indian empire, assesses that out of a total of 359 prioritized wild medical plant species assessed so far, 335 (93 percent) have been assigned Red List status, ranging from critically endangered, endangered, and vulnerable to near threatened.[54] While this government initiative is applauded, one should not forget that the inherent value of each medical plant is its unique inherent chemical properties that may contribute to fighting current and future diseases. These inherent properties of plants come from the specific nature of soil, temperature, and the adjoining physical environment. It must be established that any relocation is carried out without destroying these plants' unique medicinal properties, and such relocation cannot be called a success until the safe relocation is confirmed at individual plant level.

Two recent discoveries reconfirm the importance of biodiversity in providing some clues and answers to our complex medical and physiological needs. The first discovery is an adhesive secreted by a marine worm that inspires a promising new treatment for compound fractures of human bones. The second discovery is the electric eel's ability to generate powerful shocks, which have inspired the development of a new type of battery.[55] Both are marine animals, and sea temperature change induced by climate change could easily threaten their existence. Similarly, the US National Cancer Institute and the US Department of Agriculture found that the bark of a particular stunted tree contains molecule that are a most effective agent in treating ovarian cancer, one of the hardest cancers to treat.

Cardiovascular and respiratory illnesses, malaria, other mosquito and rodent-borne diseases, and water-borne health problems are linked to climate change. Evidence has emerged that even the El Nino/La Nina cycle affects disease. In Bangladesh the IPCC found vibrio cholerae, which is responsible for cholera, resides in algae that inhabit coastal and estuarine waters. Such algae proliferate with higher sea surface temperatures, thereby offering an increased habitat for vibrio cholerae.[56] Even without climate change algae, proliferation is a seasonal phenomenon in many areas in the subcontinent.[57] Lack of basic facilities in rural areas forces people to use moribund water from infested sources, which are a serious health hazard.[58]

Scientists also confirm that an increase in CO_2 has led to more vigorous growth of ragweed plants, which start pollinating sooner, which, in turn, has led to an enormous increase in asthma patients.

The World Health Organization (WHO) and the IPCC concur that about seven million deaths could be attributed to the climate change phenomenon. While a significant proportion of these deaths will occur in developing and poverty-stricken nations, the consequences for developed countries such as the United States cannot be ignored. It was claimed that every year a trillion gallons of sewage-contaminated water overflow from rainstorms in the United States. The situation is worse in other poorer countries such as Haiti and Bangladesh. Much of agricultural areas will suffer in these countries, with an increase in the intensity of rainfall and pollutants from these overflows.[59]

The WHO finds that the parasitic disease malaria exists in ninety countries, and India (read the entire subcontinent) is one of them. It kills over a million people a year and is second only to tuberculosis in its impact on global health. Besides human suffering, the economic impact of the disease is immense.[60] The WHO estimates that a staggering 88 percent of diseases suffered by the weaker and poorer sections of the global community are preventable through better access to water supply, adequate sanitation facilities, and better hygiene practices.[61] Unfortunately, health infrastructure is not only poor in the subcontinent; and even, if it existed, the administration is often nonresponsive to the priority needs of these communities.

Malaria has reappeared with a vengeance in recent decades. Experts believe that in the climate change context we need to understand better how the dynamics of these diseases depend on temperatures, and how the variability of temperature is going to change. They claim that lack of data remains a major block in predicting where malaria is going to occur and what its dynamics will be over time.[62] The increase in malaria and associated diseases, such as dengue fever and sleeping sickness are the direct result of temperature rise caused by climate change, yet the global response to handle the problem is somewhat muted.

Many of these diseases transcend geographical limits. It is now clear that with the changing global physical, economic and lifestyle environment (including the movement of people), malaria can spread into areas that were previously immune. For example, in Europe, tiger mosquito–induced malaria is now spreading to Germany's Rhineland and beyond. It first appeared in Switzerland, which possibly received it from tourists or movement of goods. The regional impact of a cholera outbreak in Zimbabwe affected South Africa and other border states in three sectors, namely the economy, health, and law and order.

Well-governed and resourceful states can manage these changing situations relatively easily, but the same is not true for developing nations. They suffer from poor governance and infrastructure, nutrition deficiencies, and the absence of basic medical facilities and resources. If communities (and poorer households) cannot respond to the health threats posed by climate change, their economic development will be stunted, and inter- and intracountry inequality will increase causing misery.[63] Leaving the poor and underprivileged vulnerable to new strains of diseases is economically costly and socially undesirable, as is clear from the recent HINI influenza epidemic in Mexico that rapidly spread to the United States and Canada, and then internationally through people movement.

Climate change introduces new elements that change the clinical and therapeutic characters of existing diseases, mostly requiring new approaches to treat them. The possibility of disease mutation is high because the sources of many common diseases are birds, animals (such as pigs), and insects (such as mosquitoes and flies), each of which will react differently to climate change. Many animal viruses often mutate with humans, becoming more virulent and even deadly.

According to the WHO, climate change–related deaths over the last thirty years numbered 150,000 annually.[64] This large number of deaths is taking place at a time when the full impact of climate change is yet to take effect. According to one estimate, preventable deaths from one climate change–related illness, air pollution, could be eight million by the year 2020.[65] This shows the nature of complex health issues that are associated with climate change phenomena.

Broadly speaking, climate change–induced effects fall into three categories. First is the direct impact on individuals, such as heat-related illnesses and deaths. A significant number of deaths in Europe and North America have been due to the unusually high temperature in the last few summers. Victims were mostly the aged, children, and the malnourished. A WHO expert in a televised interview in December 2009 confirmed that 70,000 more people had died by the end of the 2003 Western European summer than WHO would have expected to die in an ordinary summer. The expert claimed that this changed everything within Western Europe because WHO and the authorities in the affected countries realized that those countries were no longer immune to the effects of weather and climate.[66] Such high temperatures occur every year in the Indian subcontinent, and the victims are usually the poorest in the community. Further rises in temperature will only increase the number of deaths, particularly with increases in the number of malnourished people.

Second, changes in rainfall patterns create dangerous aquatic organisms, and cause mosquito- and snail-led fatal diseases (such as the chronic, debilitating and potentially lethal tropical disease known as bilharzia or schistosomiasis) in humans. Experience shows that people remain sick for up to forty-two days each year from such diseases.[67] The World Bank has warned India of the strong possibility that such water-related diseases will gain a foothold in that country, with disastrous health consequences.[68] These are directly attributable to a mixture of rising temperatures, contaminated water, poor sanitation, waterlogging, and the like.[69]

The third category of illnesses comes from a combined impact of complex events which in many instances evolve at a time when climate or weather patterns change. Very few people, including scientists, are able to recognize it. However, over the last half century scientists have battled against a range of deadly new diseases, such as Ebola (since 1976), Nipah (since 1991), Hendra (since 1994) and Melaka. They suspect that these deadly viruses may be at war with other species, including humans, as their habitats and sources of food vanish due to human activity.

Since 1998, more than one hundred people have died in Malaysia from previously unknown diseases. Scientists believe that the Nipah virus (originally transmitted from flying foxes to pigs and then to humans) is responsible for these deaths, and this virus is now spreading into Bangladesh and India. They believe that there is a chance, however slim, that a strain of such a virus might spark a pandemic or fall into the wrong hands.[70] The spread of HINI (or "swine flu") is one example. Most concerning is that if people in the subcontinent fall victim to such new diseases, they will often be unable to recognize it, and the spread of these new strains will have disastrous consequences for entire communities.

At a WHO-initiated conference on pandemic influenza responses, Dr. Julie Hall, senior technical adviser at the UN System Influenza Coordination argued,

> We know that . . . where you have high density, lots of people living very closely with domestic animals and wild animals, that's where you are more likely to get diseases emerging in animals and then jumping into [the] human species. We saw from the pandemic influenza this year that these types of threats really can come from anywhere in the world and occur at any time. So Asia in some ways has the disadvantage of those large and growing human populations.[71]

In countries where even basic minimum primary health care and medical facilities are not readily available to major sections of the community, climate change–induced diseases will further weaken the capacity of the existing health infrastructure to cope. This will add to the frustration of the deprived population and contribute to an already poor law and order environment. In a worst-case scenario, it becomes a security concern.[72]

More importantly, it raises a basic question of, first, access to basic medical facilities and, second, how best the health policy relates to overall development objectives, particularly when resources fall short of total needs. The approach to health policy has to be both preventive and curative in an environmental context. Countries achieving rapid development have done so by constructing their health policies along these lines.

High global morbidity and mortality rates of water-related diseases confirm this policy position. In the early 1990s, these rates were 3.3 billion and

5.3 million, respectively.[73] Diarrhea alone now causes a little more than 4 percent of all deaths and 5 percent of health loss to disability every year. Diarrheal diseases form the bulk of the health risk, with some 4 billion cases each year that result in 1.1 million to 2.2 million deaths, and DALYs (Disability Adjusted Life Years) of between 38 million and 76.3 million. Ninety percent of these deaths are children under five years of age. Adding the impacts from other water-associated diseases (such as schistosomiasis, trachoma (an infectious eye disease, and the leading cause of infectious blindness), and helminth (or intestinal parasitic worm infections), the total estimate of water, sanitation and hygiene-related ill health is 2.213 million deaths and 82.196 million DALYs each year, not including the impact from malaria.[74]

The poorer sections of the most vulnerable communities usually cannot afford medical costs, yet require a living environment that has clean potable water, a reliable sewage system, and protection from common preventable diseases through immunization. A decent public health policy should, therefore, focus on a hygienic living environment, a universal immunization program against most common diseases, and a nutritional program that builds up individual resistance against common ailments.

Modern medical treatments demand large capital investment and high operational expenses. Cost pressures, innovations, demographics, new delivery models, and rising expectations are transforming the very nature of global health care services.[75] As a result, even the United States has a very large section of the population with little capacity to seek medical assistance when they need it. Consequently, the health sector fails to meet the medical needs of those who cannot afford to pay. Although new delivery models require large capital outlays, innovative management practices may make it possible to reduce high costs and produce better health outcomes.[76]

The situation in developing countries is worse. The only option for them is to follow a different path, and preventive care is certainly a better option along with improving the functioning of the institutions to carry out social development more efficiently.[77] When focus is on preventive care, payers will be able to align reimbursement plans to support such care and effective treatments without looming over the shoulders of providers and patients. On ideological grounds, some may find it difficult to accept this position. However, at the same time, it must be remembered that in a gradually depoliticizing society, political ideology is playing a less central role in defining social problems and their solutions. Further, marketization has also reduced the capacity of governments to control societal resources, and, in a way, even discouraged government intervention.[78]

The WHO is unequivocal that a very significant amount of disease in developing countries can be prevented through better access to safe water supply, adequate sanitation, and better hygiene practices by focusing on

- environmental health (such as a functioning sewage system);
- universal immunization against most preventable diseases; and
- meeting all basic nutritional needs (particularly those of vulnerable groups).[79]

In 2006, global expenditure on health was 7.6 percent of the global Gross Domestic Product (GDP). Health expenditure is also a major element in determining the UN's Human Development Index (HDI). Detailed data indicate that, in some cases, major differences exist between public and private health expenditure. In 2004, the health expenditure of 15 countries was above 9 percent of their GDP, and their HDI ranking was within the top 40 countries. These countries are the United States (15.4 percent), followed by Switzerland (11.5 percent), Germany (10.6 percent), France (10.5 percent), and Austria (10.3 percent). Compared to high HDI countries, health expenditure in subcontinental countries is miniscule (table 4.2).

However, when expenditure between the public and private is considered, one gets a different picture. The top five public expenditure countries are Iceland (8.3 percent of GDP), followed by France and Germany (8.2 percent each), Norway (8.1 percent), and Austria (7.8 percent). The picture is different when private expenditure is considered. These countries then are the United States (8.5 percent), followed by Argentina (5.3 percent), Switzerland (4.8 percent), Uruguay (4.6 percent), and the Netherlands (3.5 percent).[80]

The benefits of high expenditure in the health sector are visible in the UN Statistics Division's Life Expectancy report: the Organization for Economic Co-operation and Development (OECD) (0.88 percent), least developed

Table 4.2. Public and private health expenditure in selected countries in 2004

Country	*Public expenditure: percentage of GDP*	*Private expenditure: percentage of GDP*	*Per capita US$ expenditure (purchasing power parity)*
Cuba	5.5	0.8	229
China	1.8	2.9	227
Bangladesh	0.9	2.2	64
Bhutan	3.0	1.6	93
India	0.9	4.1	91
Nepal	1.5	4.1	71
Pakistan	0.4	1.8	48

Source: United Nations Development Programme 2007/2008: 248–49.

countries (0.45 percent), and South Asia (0.64 percent).[81] In South Asia, it varies between 0.8 percent for Bangladesh, 1.4 percent for Nepal, and 1.3 percent for India.[82]

The table speaks for itself, and certainly provides a poor picture for most subcontinental countries. These countries have little capacity and limited financial resources to tackle most climate change–induced diseases.

A critical strategic issue is how these countries maximize their returns from their limited health budgets. The world's poorest people spend on average 50–80 percent of their household income on food,[83] and the largest numbers of these people live in South Asia. In such circumstances, good governance practice demands that the neediest have access to a health safety net. It would be cost-effective and achieve other social and economic objectives at the same time. More importantly, experience shows that the return on environmental and preventive investment is significant.

The governance issue becomes a critical adjunct for the success of any health policy. A health policy cannot be effective if applied in isolation; its success is intimately connected with other public policies. Three strands of health policy are environmental, preventive and immediate medical needs.

Most health professionals acknowledge that the cost-effective way of providing best health facilities in a resource-constraint environment is to focus on the environmental and preventive aspects. Responsive health facilities demand creating treatment facilities for those medical conditions that fall into all three categories. Like all public policies, health policy must gel with other national policy objectives: it has to be a total approach that includes gains in productive time, treatment cost reduction, and the value of prevented deaths.

Health experts generally agree that, of all the developing countries, Cuba's health policy meets the basic needs of its people. Its basic focus on health policy has been equity and priority of care for vulnerable groups, and its policy emphasis has been on total health management.[84] This policy was implemented by merging clinical practice and public health with the national health care system.[85]

Cuba's public outlay on health has been 5.5 percent of GDP. Its total per capita expenditure on health in 2006 was US$229, compared with US$6,096 for the United States. Many consider that the US health care system to be inefficient.[86] US president Barack Obama's unyielding crusade to reform the US health sector confirms this position. The United States outranks Cuba by only two positions (37 to 39, respectively) in the latest WHO report on global health care. This means there is little difference between the two countries.

However, it is not implied that countries should blindly follow Cuba's health policy. On the contrary, they should develop their detailed policies based on their national realities, and focus on all aspects of socioeconomic-political objectives and priorities, keeping in mind the reasons behind the success of health policies in Cuba and elsewhere. The WHO has also provided guidelines for developing-country health policies, and these also should be taken into consideration when developing their respective health policies. These are

- evidence-based prevention;
- enhanced surveillance for drinking water–associated illness;
- support of illness outbreak investigations and emergency responses; and
- epidemiologic and laboratory research on drinking water issues.

While the experience of other countries is a useful guide for the development of health policy, in the climate change context even this may be inadequate to capture the existing and emerging health and medical needs of individual countries. It is certain that when climate change has evolved further, new diseases will emerge. This will require intensive monitoring and a quick dissemination of all aspects of such diseases to all relevant professionals and policy makers so that the collective knowledge of the scientific community, health professionals, and other relevant policy personnel become aware of such developments. Only their collective knowledge can respond to the needs of the rapidly changing health environment; there is no alternative.

FOOD

Food security remains the biggest policy concern for South Asian countries, as it has the potential to threaten the security and stability of states when shortages coincide with other threatening elements.[87]

Even without the climate change phenomenon, agriculture in South Asia has failed to keep pace with increasing demands. Afghanistan's situation remains critical: it experienced rapid decline in food production during the 1990s. Bangladesh and Pakistan reversed a negative per capita production trend in the 1980s to a positive trend in the 1990s, but these trends are now tapering off. Nepal went from positive trends in the 1980s to negative trends in the 1990s. Overall, in the decade to 2000, Afghanistan, Nepal, and Pakistan had food shortfalls. Between 1991 and 2000, these countries received substantial food aid to meet their food shortfalls.

India's food production trends were positive in the 1980s and 1990s, but it has since slowed. In the recent years, like other subcontinental countries, India is also importing food.[88] The 2007–2008 global food crisis sent wheat and rice prices soaring to record highs, sparking hoarding and riots. Commentators considered this to be a wake-up call.

It is estimated that world food output must grow by 70 percent over the next four decades to feed a projected extra 2.3 billion people by 2050. Commentators argue that the last time the world faced a food crisis of this magnitude was in late 1960s and early 1970s. Donald Coaxe, a portfolio strategist forecast in 2008 that a global food catastrophe would reach further and be more crippling than anything the world has ever seen. He specifically referred to the demands by India and China.[89] It is worth noting that the aggregate population of China, India, Indonesia, and Bangladesh in 2004 was 43.2 percent of the global population.

India and China transformed their agricultural system to become self-sufficient in food,[90] but the situation in India has been changing fast. For example, the Punjab (the cradle of the first green revolution), in a letter to the Indian Ministry of Agriculture has expressed its inability to take on further responsibility because its groundwater level is rapidly receding.

The Punjab claimed that its water table decreased by an average of 10 m between 1999 and 2009. It stood at 12.80 m in 1999, and dipped 29.40 m in 2009. It has even been as low as 30 m in central Punjab. However, the Indian government did not mention land salinity, the other critical issue. Both groundwater level and land salinity issues are extremely difficult to address in the short to medium term. Remedial policy initiatives require a large injection of capital. In other words, the situation needs a multipronged policy initiative based on priority.

The Indian government appears to be clueless on what to do, even in the context of recent unprecedented rise in food prices in the country.[91] Its Minister for Consumer Affairs, Food and Public Distribution is reported to have sent the draft food security bill to the prime minister's office for formal clearance. According to the media report, the quantum has been fixed in the draft at 25 kg per month per head to the vulnerable section of the community at the subsidized rate, although the Supreme Court directed it should be 35 kg. There remain glaring gaps between the judge's decision about the quantum of food required and the government's decision, should the prime minister agree with the draft recommendations.[92]

It has been known since 2004 that a rise in temperature leads to a decline in rice productivity.[93] Researchers elsewhere were also advising that warmer temperatures adversely impact on rice productivity. It is only recently that this issue received the serious attention of the Indian government's agencies.[94] A forward-looking policy should have addressed this issue long ago, because food deficit has the potential to cause economic damage and may even threaten internal security. Rising grain prices have led to riots and mass demonstrations in Indonesia, Mexico, Haiti, and Italy.[95] Police faced violent demonstrations in Senegal and other rice-dependent nations in West Africa. Even the largest US grocery chain restricted rice sales to customers.

India's failure to seriously understand the reasons for its declining food productivity is a serious policy failure, as research takes a long time to produce results. Three significant problem areas for India are lack of water at the right time of the crop production cycle, land salinity, and temperature rise caused by climate change.

Demands by India, China, Brazil, Indonesia, Bangladesh, Pakistan, and Nigeria are a critical driving force in determining the international price for cereal. Collectively, they constitute close to one-half of the global population. Although Africa did not emulate the successes of larger countries, recent indications are that some African countries have been able to turn the corner.

The Washington, DC–based International Food Policy Research Institute (IFPRI) considers China's rice program to be one of the real success stories in agricultural development. China raised its yield from two tons per hectare in the 1960s to more than 10 tons by 2004. It has now established a target of 13.5 tons by 2015. Successes achieved in India might have received high acclaim, but the water levels in India are drying up, and its soil has been degraded by pesticide, an excessive use of fertilizer, and land salinity. Climate change is creating a new menace, as it encourages an increased level of pests, and plant diseases.

The IFPRI argues that climate change–induced lower yields would cut calorie availability for the average consumer in developing countries such as India by 7 percent over the next four decades, compared with 2000. It further argued that South Asia would experience the largest decline in yields in almost all crops, and that, even without climate change, rice output would decline by 14 percent.

The global population is increasing by about seventy-nine million each year, of which India, China, Nigeria, Pakistan, Indonesia, and Bangladesh account for about half. India represents about 21 percent of this growth. Only China's fertility level of 1.8 births per woman is below the replacement level.[96] Any extra food requirements for these countries would add to existing global food shortages. Assuming a minimum per capita consumption at 1997–1999 levels, the additional food requirement is about 13 million tons by 2010. Besides this, the shortfall to meet the nutritional gap by 2010 is estimated to be 22 million tons.[97]

The combined population of India, Bangladesh, Nepal, and Pakistan in 2004 was 1,397.5 million.[98] China's population growth rate will stabilize in the near future, but other countries will take a long time to achieve this. In other words, all else being equal, these countries will continue to experience food and nutritional deficiency. In 2010, forty-two countries face food deficit and sixty-seven countries experience nutrition deficiency, but, because of the large population base in these countries, their additional food requirements will be immense. Rice and wheat dominate the basic food intake in the subcontinent, and the water requirements for rice cultivation is nearly double that of the equivalent amount of wheat.

A large number of people below the poverty line make policy tasks more complex (table 4.3). The 2008–2009 global financial crises only compounded the problem.

In the subcontinent, income distribution, generally speaking, favors the rich; with major food policy implications. The food habits of the affluent class differ from those of the poorer in the community. Their dietary habits change in favor of high-protein foods such as meat and dairy products, including poultry. This has become the situation in India.[99] Demand for meat in China quadrupled over the last three decades. The conversion rate of grain to meat is low; about 7 kg of grain is required to produce 1 kg of meat, while 40 percent of poultry value usually reflects the grain price.[100]

Rice is eaten in its various forms by three billion people in Asia. Only in Japan has rice consumption been declining (since the 1960s), to be increasingly replaced by bread and meat, although it remains an important element in Japan's culture and religion (Shintoism).[101] Notwithstanding, the average per head annual rice consumption is about 60 kg, or one-half of the consumption level prior to the Second World War. Further, Japan now imports much of its rice.

Rice is also one of the least internationally traded commodities. Even in normal times, more than 90 percent of the production is consumed where it is grown.[102] Producing countries traded only 0.28 percent of their produce internationally in 2003–2004. China, India, Indonesia, and Bangladesh produced 76.21 percent of the total global rice production in 2003–2004 (table 4.4).

Table 4.3. Percentage of population living below the income poverty line

Country	*US$1.25 a day (2000–2006)*	*US$2.00 a day (2000–2006)*	*National poverty line (2000–2007)*
Afghanistan	. . .	. . .	42.0
Bangladesh	49.6	81.3	49.8
Bhutan	26.2	49.5	. . .
India	41.6	75.6	33.0
Nepal	55.1	77.6	30.9
Pakistan	22.6	60.3	32.6

Source: United Nations 2008: table 3.

Table 4.4. World's major countries producing, exporting, and consuming rice, 2003–2004 (in '000 metric tons)

Country	*Production*	*Export*	*Consumption per 1 million population*
China	118,000	2,500	103.344
India	89,000	3,000	78.916
Indonesia	33,000	. . .	152.702
Bangladesh	26,000	. . .	182.927
Vietnam	21,000	4,000	211.885
Thailand	17,800	8,000	158.913
Myanmar	10,440	500	217.035
Philippines	8,500	. . .	110.407
Brazil	7,300	. . .	43.522
Japan	7,100	. . .	67.950
United States	6,174	2,900	13.127
Pakistan	4,500	1,600	. . .
South Korea	4,500	. . .	103.123
Egypt	3,900	700	42.577
Taiwan	1,140	. . .	50.2315
Australia	358	250	. . .

Source: Data from the US Department of Agriculture, http://www.nationmaster.com (accessed February 3, 2009).

Other countries produced the remaining 23.79 percent.[103] This implies that, even when only one major producing country experiences a production shortfall, the global supply-demand equation alters, leading to price increases in the global market. For example, Australia has been a rice production and export country, but its production and exports have already declined by about one-third, which is primarily due to water shortages.

Currently Thailand, Vietnam, and the United States account for sixteen million of the twenty-five million tons of world rice exports.[104] Of these three, Thailand is the only country with a surplus (about 4 million tons).[105] During the height of the crisis in 2004, China could not get the expected 500,000 tons of rice from Vietnam when Vietnam restricted its rice exports.

Although wheat is not an exact substitute, most Indians are familiar with it. It is reasonable to assume that when rice production is affected because of climatic conditions, wheat production may also face the same situation. As India, Pakistan, and China are among the world's major wheat producers (table 4.5), the food situation may be less desperate from any decline in rice production. However, Bangladesh could find itself in a slightly uncomfortable position, as traditionally wheat is not a part of its food culture and habit.

Table 4.5. World's top ten wheat-producing countries in 2005

Country	*Amount (in million metric tons)*
China	96
India	72
United States	57
Russia	46
France	37
Canada	26
Australia	24
Germany	24
Pakistan	22
Turkey	21

Source: Data from the US Department of Agriculture, http://www.nationmaster.com (accessed March 27, 2009).

There are two broad options available to overcome internal food shortages: increase domestic production, or import from outside.

Additional production requires improved productivity or bringing additional land under food crops. Besides increasing capital input to improve land productivity, land reform is an essential precondition for improving productivity in the subcontinental countries. Even the World Bank acknowledged more than a decade ago that tenancy reform and effective implementation of land ceilings is critical for improved productivity. Such processes have been successfully introduced only in Kerala, West Bengal and Karnataka in southern India, South Asia.

Increasing rice productivity is complex. Most of the world's increase in rice production over the last twenty-five years has occurred close to sea level in river deltas or fertile river basins. Researchers, including those based at the International Rice Research Institute (IRRI) in the Philippines, have found that those rice plants which bloom on hot days are less likely to produce grains of rice—a difficulty already starting to emerge in inland areas of China and other Asian countries as temperatures climb.[106]

Subcontinental countries do not have any significant scope to bring additional land under cultivation, because population density is already high in these countries (table 4.1). Further, land required for urbanization, industry, industrial raw material, and recreational purposes is forcing the conversion of food-producing land to other purposes. Even in China, a one-party state, farmland losses are approaching crisis levels.[107] China's grain area fell from 90 million hectares in 1999 to 77 million hectares in 2004. Arresting the shrinkage depends on halting the expansion of deserts and controlling the conversion of land currently in grain to nonfarm uses. It is an extremely costly exercise. For example, it is estimated that desert remediation in China alone will require an investment of some US$28 billion.[108] Shielding cropland from nonfarm demands is politically complex, as recent events in India and Bangladesh confirm.

Of the five subcontinental countries, Bangladesh has the highest rural population density. Most of south and central Bangladesh faces the possibility of inundation by rising sea levels. The rate of displacement will dramatically increase, as the impact from rising global temperatures take full effect. Nepal's terrain is hilly, with only 17.5 percent of its land being suitable for agricultural purposes. There is little scope for using extra land for food production.

In 1990 about one-half of India's land area was salt affected.[109] Satellite imagery of the Ganges plains in Haryana and Uttar Pradesh is already showing the symptoms of "salt-infested" land.[110] The United Nations Economic and Social Commission for Asia and the Pacific (ESCAP) estimated that salinity was less of an issue for Pakistan than India, although S. Aziz, an expert in Pakistan on the subject, warned of the dangers of salinity decades ago.[111] The situation in Bangladesh is different, as annual flooding compensates for such losses.[112] Topsoil loss in India has already surpassed tolerable limits. It is India's most precious resource, and an inch of topsoil takes roughly one thousand years to form.[113] Aziz argued that the absence of an integrated approach was causing irreparable damage to long-term agricultural productivity. These are warning signs that food production will decrease irrespective of the effects of climate change.

Another key requirement for any productivity increase is the availability of water at the right time of the production cycle. Water shortage is already a serious barrier to the existing landholders and will remain so even if new areas can be brought under cultivation. A key point is that, even if sufficient water were available, cereal production in developing countries would still be less than half the amount produced by developed countries in the 1990s.[114] To give an extreme example, the US state of Illinois produces more than two hundred corn bushels per acre compared with an average of thirty bushels per acre elsewhere in the world.[115] There are, of course, other factors that are critical for improved productivity.

It is estimated that, at current rates, 2,000 km^3 of additional irrigation water is needed if the world's population is to be fed by, for example, 2025. As agricultural operations are already creating water deficits,[116] questions remain about the source of any extra water. Available data indicate that currently water used to produce about 10 percent of the globally harvested grain is not replenished.

Regardless of climate change, water is an extremely precious natural resource in many countries. Increasingly, most countries are introducing an economic price for water that has led to increased costs for rice production. For example, even in Australia with the recent doubling of rice prices to around US$1,050 a ton for high grades of rice, it is more profitable for farmers to grow grapes for wine production, as these produce a pretax profit of around US$5,200 per hectare compared to US$620 for rice. Many rice farmers have already sold their water rights.

FISH: AN IMPORTANT SOURCE FOR FIGHTING NUTRITIONAL DEFICIENCY

Rice, wheat, corn and similar products meet the staple food needs of most economically poor communities in the subcontinent. Usually peoples' nutritional needs demand that they should have some access to protein at an affordable price. Unfortunately, a large section of the subcontinental economically disadvantaged population has minimal access to protein on a regular basis. For South Asian countries and China, marine and aquatic products are increasingly becoming an important protein source, because of the need to provide protein to nutritionally poor people and the upper middle class.

Over the centuries, village ponds, streams, lakes, rivers, and seas have provided some protein to the underprivileged. These sources are rapidly drying up as a source of such protein. Increasingly, most inland fishing grounds have become barren due to industrial and commercial pollution, the construction of dams in rivers upstream, and overfishing. Coastal areas also no longer provide fish, as many countries, with the help of modern fishing technology, have left coastal areas with depleted fish stocks. Climate change and chemical pollution (including from fertilizer runoff) is also responsible for the destruction of many fishing grounds.

Seas are the largest ecosystem on the planet. No individual country can address these problems unilaterally. Global governance can only go so far, by adopting initiatives such as restricting open access to fisheries, taking a tougher stand on illegal fishing estimated to be worth US$10 billion to US$23 billion per year, and compulsorily limiting the quotas of fishing fleets. After the Second World War, acoustic and navigational technologies allowed countries to track fish.

This bounty from the seas has steadily declined over the last few decades. Even marine-protected areas cannot be successful if they have lost the capacity to provide fish with food and a home.[117] Despite this, few resources have been committed to understanding the ecological complexity of these underwater worlds.[118] Further, fishing methods have become more destructive, contributing to overfishing and damage of marine ecosystems. Worse is the lack of any political commitment by governments to tackle the problems. Seas are thus being quickly depleted of their fish through wasteful fishing methods and illegal fishing.[119]

In Asia, nations such as Japan, South Korea, and Taiwan are guilty of such practices. In particular, Japan, a large fish consumer, takes an intransigent position in this regard. Japan and several other nations have tried to prevent any global initiative to restrict first, unsustainable and unrestricted catching practices; and, second, any measures that would strengthen Convention on International Trade in Endangered Species (CITES) of wild fauna and flora, even though they may not be involved in catching, certain varieties of fish such as the Atlantic blue fin tuna.[120]

As fish stocks vanish from their natural habitats, people are increasingly focused on inland aquaculture. In 2005, just over 30 percent of the global fish supply came from fish farms and this percentage is on the increase. Experience to date indicates that the growth rates of only two types of fish—carp family and the tilapia family—looked promising. Unfortunately, they provide little nutritional value. Carp are omnivorous and, although they can eat a vegetarian diet of water plants, they prefer to scavenge the bottom for insects, crustaceans, and benthic worms. Tilapia fish, meanwhile, eat insect larvae, shellfish young fish, worms, plants and any minor edible waste. The Food and Agriculture Organization (FAO) has provided a grim picture of the global fish supply situation:

- fifty-two percent of global fish stocks are fully exploited;
- another 24 percent of global fish stocks are overexploited; and
- seven of the top marine species (accounting for 30 percent of all production) are fully exploited or overexploited.

Fishery experts agree that the consequences of climate change are devastating for fish and plant survival.[121] It is virtually impossible to overcome fish shortages by fish farms alone. It is predicted that the combined impact from climate change and decreased or stagnant water flow may result in losses between 4 percent and 22 percent of native fish biodiversity in over one hundred rivers. Losses in individual rivers could be as high as 75 percent. Rivers experiencing increased flow also suffer unwanted ecological change due to the invasion of nonnative species that are better acclimatized to higher flows.[122]

As more than one-half of global fish stocks have already been exploited, and a few countries with access to modern technology are harvesting fish by scooping fish from every possible corner of the sea floor, it is almost certain that the situation will worsen further in the near to medium future, unless radical changes are made to existing fishing practices and related governance issues.[123] Past experience indicates that laws may be passed and protocols developed, but implementation and monitoring remain wanting. The perpetrators of illegal practices always find ways to subvert such laws and practices, even though it may involve corrupt practices.

Fish is an important dietary item and source of protein for the inhabitants of eastern India and Bangladesh. A large proportion of their total demand used to be met by fish harvested from (as noted above) streams, rivers, lakes, and ponds. In eastern India and Bangladesh fish harvests have declined significantly, mostly caused by population pressure and increased pollution levels in fish habitats. More importantly, a large variety of fish can no longer be found in the region. In other words, biodiversity has been the biggest casualty, with some species being lost forever.

This is confirmed by the Environmental Performance Index produced by a team of environmental experts at Yale and Columbia universities, which ranks 163 countries on twenty-five performance indicators tracked across ten policy categories.[124] The index shows that of those countries, India had a ranking of 123 (with an index score of 48.3). Other index scores for the region were Bangladesh (44.8), Pakistan (48.8), Nepal (68.2), and Bhutan (68.8). Of most concern is China, which also has a very poor index value. Many rivers, including two of India's perennial rivers—the Indus and the Brahmaputra—originate in China. There is a high probability that these two rivers carry some pollutant elements from China, an outcome that is of serious concern to the fishing industries of lower riparian countries.

Researchers now argue that the widespread extinction of ocean floor organisms during the Palaeocene-Eocene Thermal Maximum, which saw greenhouse warming and acidification, indicates that similar future extinctions are possible due to climate change (with the attendant growth of poisonous algae and decline in fish stocks). Harvesting fish from the seas in large amounts does not help the situation.[125]

RISING FOOD PRICES

Estimates of global food price inflation vary. One estimate puts the average increase (since early 2006) in the global price of rice at 217 percent and for wheat at 136 percent. Both the United Nations World Food Programme (WFP) and the IRRI confirm that food prices worsened globally during the 2007–2008 period and that this rise is continuing. In the Asia-Pacific region, the cost of rice is now at a twenty-year high after prices rose by about 50 percent in 2007. Food price inflation in India had risen by nearly 20 percent by the end of 2009, and the situation has reached crisis level (e.g., sugar prices rose by about 60 percent, pulses by 46 percent, and even potato by 53 percent over that same year).[126]

Indeed, the subcontinent saw the rice price double in the space of a few months in 2008.[127] In 2004 a typical poor farmer in India was spending more than half their daily income on food, and that was before grain prices increased dramatically.[128] In January 2010 the inflation rate declined only marginally.[129]

The governments of subcontinental countries try to overcome food inflation situations by subsidizing the price of staple foods (such as rice) to selected sections of the community. However, such initiatives are open to corruption, and their financial cost is high. More importantly, given weak fiscal and revenue bases and other priority needs, it is arguable whether governments can continue to subsidize the price of essential commodities when there are substantial increases in international prices.

Most South Asian countries provide subsidies for basic staple foods to targeted sections. The recent increases in rice and wheat prices have put a major dent in the budgetary positions of both the Indian and Pakistan governments. In order to maintain the subsidized food prices for the poorer sections (at about two-thirds of the domestic market retail prices), India has been forced to raise its budget allocation for food subsidies to around US$8.3 billion compared with US$6.0 billion in 2006–2007. Pakistan subsidized a food ration scheme at a 42 percent discount of market price. The cost of the scheme forced the government to initiate major cuts in expenditure in other priority areas.[130] Given their uncertain fiscal state neither country can afford to subsidize food indefinitely.[131]

Sometimes political considerations require the adoption of such policy. However, this option should only be considered as a last resort; it is not a panacea for every time there is a shortage. Governments need to make sure that subsidies are provided only to the genuinely needy, and that subsidy payments are not continued in perpetuity. Administratively, good governance practice also requires that the system does not become a fertile ground for generating corrupt practices, which, unfortunately, is usually the case in the subcontinent.

FOOD VERSUS BIOFUEL

Notwithstanding our ingenuity and technological progress, the basic food for all nonaquatic animals is produced, primarily on land. The physical supply of this land remains fixed. Yet demand for it is increasing exponentially with a growing world population. Increase in land productivity is certainly one option to offset this, but without significant investment in infrastructure and new investment any productivity increase is unlikely. Generally speaking, climate change makes the task more complex, and more so for the governments of developing countries.

Agricultural scientists predict that by 2050 agricultural productivity will have declined, and that the range of loss will be 1 to 20 percent. In Canada, it will only decline by 1 percent, while India could face a maximum decline of 25 percent. Besides India, two other countries most affected by loss of agricultural productivity are Brazil and Australia (with reductions of 10 and 17 percent, respectively). All things being equal scientists estimate that rice yields will decline by 15 percent for every 1°C increase in daily mean temperature.[132] This suggests that global rice yields could fall by about one-half.[133]

While the need for additional food is increasing, there is also active pressure to cultivate biofuel-producing crops. In a competition between rice and olive farming (required to produce wine and oil), olive wins because of its potential to provide higher financial return to the grower. As biofuel demand increases, it is almost certain that growers in increasing numbers will opt for cultivating a biofuel-producing crop, replacing their low-value cereal crop wherever conditions permit.

The steep rise in fuel price during 2007 and growing public concern to reduce fossil fuel gas emissions have largely contributed to the push for biofuel production. The United States is at the forefront of this policy. Its policy is to produce 133 billion litres of nonfossil transport fuel by 2017.[134] Some US states give tax breaks to companies using biofuel, and analysts expect the demand for such incentives to grow. The European Union has also set a target of a 20 percent replacement of fossil fuels with biofuel by 2020. The shift toward biofuel has devastating consequences for rain forests and food supply.

The consequences are far-reaching. As well as raising the basic price of food, it has encouraged the felling of tropical forests. Politicians in power and those with vested interests have not slowed their vocal desire for biofuel.[135] Biofuel proponents claim that using ethanol would bring a 20 percent saving in greenhouse gas emissions compared to petrol.[136] However, a team of independent interdisciplinary research scientists provide a different conclusion. They argue that "corn-based ethanol, instead of producing a 20 percent savings, nearly doubles greenhouse emissions over 30 years and increases greenhouse gases for 167 years. Bio-fuels from switch grass, if grown on US corn lands, increases emission by 50 percent. This result raises concerns about [any] large bio-fuels mandate."[137]

Besides this desire for ethanol production, demands for animal food, raw materials for industry, manufacture of other food products, and cosmetics are also increasing at an exponential rate. Europe, the United States, China, and

Japan are all hungry for energy. They are asking farmers to grow more maize, sugarcane, oil palm, and oil seed rape. John Vidal argues that the outcome of advocacy for biofuel is not only that "the era of cheap food is over," but that such an option will hit "the poor and the environment the hardest."[138]

This is confirmed by other studies. William Cline confirms this in his country-by-country assessment of the impact of climate change on agriculture.[139] He concludes that India is especially at risk, facing agricultural output losses of nearly 30 percent even under the best scenario, with a decline of as much as 44 percent in the northeast if there is a lack of carbon fertilization.

As international pressure mounts to combat greenhouse gas emissions, experience indicates that rich countries will try to buy out of their difficulties by offering money (with no strings attached) for products that replace fossil fuel. The rich are unlikely to sacrifice their lifestyle, and will opt for adaptations rather than profound solutions.[140]

The IFPRI has highlighted the immediate consequence of price rises and this push for biofuel. It projected in 2003 that the number of global hungry people would decline to 625 million by 2025, so long as the relative price of food remained constant. However, the policy push for biofuel suggests that the number of food-insecure people in the world will rise by over 16 million for every percentage increase in the real price of staple food. This means that 1.2 billion people could be chronically hungry by 2025, 600 million more than previously predicted. The WHO estimates that the food price crisis has set back any progress made in reducing poverty in poor areas by seven years, and that this will impact on the vulnerabilities of societies and even households,[141] as indiscriminate deforestation and peat draining make way for increased oil palm plantations and acacia for paper pulp production due to the increasing pressure for biofuel production.[142]

Similar to the deforestation of the Amazon rain forest, an alarming amount of land and swamp has been cleared in Brunei, Malaysia, and Indonesia. In 2000, Indonesia had 4.2 million hectares of land under oil palm production: in 2007 this increased to 7 million hectares. The estimated amount of land slated for production in coming years is 20 million hectares.[143]

Collectively, since 1970 Malaysia and Indonesia have drained 151,000 km^2 of peat swamps. The drained peat swamps have released 2 billion tons of carbon, an amount approximately equal to 8 percent of the annual global emissions from the burning of fossil fuels. A Dutch government–affiliated research institute, Wetlands International in collaboration with Delft Hydraulics, found that if emissions from peat swamps were factored into calculations of national greenhouse gas emissions, Indonesia would jump from twenty-first place to third, behind the United States and China.[144]

The impact of such deforestation is not necessarily limited to the immediate region, but to a much wider area, as it influences the climate through increased carbon emissions.[145] This has been happening in Southeast Asia, because of large-scale clearance in Indonesia.[146] Dr. John McCarthy, a leading researcher in the field, was forthright in his comments that oil palm is a simple plant, albeit one wrapped up in many issues.[147] A report from the UK-based Global Canopy Programme, an alliance of rain forest researchers, confirms this.[148]

Globally, the forestry sector is one of the most complex sectors of primary industry. The ecological and climatic consequences of forest clearance are far-reaching, and the damage irreversible. Policy approaches have to be different, as "forest destruction is caused by an incredibly wide variety of political, economic and other factors."[149]

Researchers at Cambridge University, the Museo Regional de Ica (in Ica, Peru), and the French Institute of Andean Studies in Lima, Peru, found that the huarango tree, a crucial element in desert's fragile ecosystem, is now gradually being replaced by crops such as cotton and maize. The huarango tree enhanced soil fertility and moisture and helped to hold the Nazca's narrow, vulnerable irrigation channels in place. The Cambridge study concluded that the demise of this South American civilization, which flourished in the valleys of Peru's southwest coast and disappeared about fifteen hundred years ago, was helped by damaging the fragile ecosystem that held it in place.[150] The ecological consequences of large-scale forest and swamp clearance remain largely unknown at this stage.

Also, a recent concern has not yet received the attention it deserves from the policymakers and climate and environmental activists. Most governments, by nature, prioritize their immediate threats. In recent years many governments, particularly those from countries in Central Africa, Southeast Asia, and South America, have paid little or no attention to land acquisitions by foreign governments or multinational companies of vast tracts of farmland. The United Nations estimates that 74 million acres of farmland in the developing world were acquired in such deals during the first half of 2009, an amount equal to half of Europe's farmland.[151] It is believed that mostly food importers, particularly those in the Persian Gulf and East Asian countries, are involved in such mechanisms.

Such food importers get their food from countries that apply fossil fuel–based technologies (including the intensive use of chemical fertilizer), use an increased amount of water, and cut down trees to clear land. Such practices have wide-ranging public policy ramifications as these are all environmentally questionable practices that have negative environmental, economic, and social consequences. Further, the food importers behind such practices are usually economically powerful and politically well connected. For example, informal reports from Cambodian farmer groups in January 2010 indicated that Phnom Penh was setting aside national regulations on forest protection and preservation so that foreign firms could convert forests into large-scale plantations.

However, the failed Copenhagen Summit did provide some glimmer of hope in this regard. In the 1997 Kyoto Protocol, the topic of forests remained submerged in an ocean of political intrigue that often took center stage in discussions. One reason for this omission was Brazil's unwillingness to accept limits on its freedom to fell trees.[152] But this changed in Copenhagen, as proforest groups ran a skillful campaign to establish their point.

The group received strong support from the United Kingdom's Prince Charles. His involvement helped the group to achieve some success in their negotiations concerning reduction and destruction of the world's forests. The United Nations focuses on this problem in its collaborative program on Reducing Emissions from Deforestation and Forest Degradation in Developing Countries (REDD).[153] Most agreed that the felling of trees contributes up to 20 percent of global emissions. During the discussions, rich and politically powerful nations such as Australia, France, Japan, Norway, the United Kingdom, and the United States endorsed tree protection in terms that went beyond the immediate need to stem emissions. They have collectively pledged a payment of US$3.5 billion, as the first installment of the US$25 billion needed by 2015.

The *Economist* argues that the Copenhagen Accord delivered at the summit is hardly far-reaching, although it is better than nothing. It applauds the funding offer for short-term projects to developing countries and the commitment of US$100 billion in the long run, and the agreement on reducing emissions from deforestation and forest degradation. Equally important is the Accord's inclusion of both developed and developing countries, which was not the case with the Kyoto Protocol.[154] Some argued for caution that although the agreement noted above offered significant positive implications for tropical forest management, many issues related to it remain unresolved. Hence, there is also a strong case for giving research priority to the study of unknown aspects, such as are noted below.[155]

To minimize the adverse affects of climate change, it is essential that all options are examined carefully to ensure optimization of return on scarce investments. Given most countries' concern for the economic consequences of climate change, it is important that both short-term and long-term policy issues are resolved as far as practicable at the outset. It appears that the initiatives of the REDD program so far have paid more attention to mitigation measures than adaptation. However, such an emphasis may reduce the longer-term ability of rural households to get out of poverty in nations where small landholders dominate the agricultural sector.[156]

There is another downside to adaptation. If an international burden-sharing agreement for adaptation financing is established, potentially a large transfer of wealth may be directed only toward the Alliance of Small Island States (AOSIS) and least developing countries, because they will experience the largest impacts from climate change over the short and long term.[157] Such issues are politically sensitive, as some donors may wish to have a larger mix of beneficiary countries. Such issues are politically sensitive and need to be addressed.

Many NGOs that were behind the successful campaign remain skeptical. The Copenhagen Summit set neither a target nor timetable for achieving the objectives in Copenhagen.[158] NGOs stressed the need for a 50 percent reduction of deforestation by 2020 and a complete halt by 2030. But forested nations were unwilling to accept this timetable until they saw the final financial offer and how any funding would be distributed. The discussion indicated that any distribution would be based on a vigorous performance audit and rigorous verifications. Yet NGOs at least can claim one definite achievement. Anglo-Dutch multinational Unilever agreed, in principle, to abide by agreement reached at the summit. It also announced that it had abandoned one of its oil palm suppliers because of doubts about what the supplier's activities in some of the world's forests.

INTERNAL AND EXTERNAL SECURITY

The strategic and security concerns of a country are manifest, and relate to both external factors and internal threats. For most states in South Asia domestic threats to security have become important during the past decades.[159] However, in a globalized world, internal issues often remain subservient to external factors. Contributing factors to these issues have been the end of the Cold War, the emergence of economic globalization as the new economic mantra in the new millennium, and the revolution in communication and transportation.

Globalization and the emergence of civil society and human rights issues have injected new dynamics into the security concerns of nations. Increasing mobility has established diasporic connections between the home and host countries. This issue is alone capable of impacting on the security concerns of a country.[160] Diasporic groups are becoming increasingly prominent players in the security concerns in their host countries, in addition to their homeland.

Examples are demand for a Tamil homeland in Sri Lanka, the Khalistan movement in India, and the Baloch secessionist movement in Pakistan. In addition, religious fundamentalism has also drastically altered the security concerns of many countries, particularly in the subcontinent. Most concerning is their decentralized decision-making process that cultivates homegrown insurgency movements, making it impossible for individual countries to preempt security threats from such sources.

In previous centuries, particularly before the beginning of the First World War in 1914, security issues had very limited connotation such as border security and maintaining internal stability. For example, in 1877, the year that Queen Victoria became Empress of India, a famine in India's south killed five million people, even as Queen Victoria's viceroy remained adamant that famine relief was a misguided policy. Similarly, some 2–3 million people died in the Bengal famine of 1943, caused by a combination of border security during the Second World War and government mismanagement, and not solely because of food shortages. Neither of these incidents raised much outcry globally, or within the nations involved.

These issues are an interesting background to the issues covered in this book. They show how perceptions change with the passing of time and the changing global sociopolitical-economic environment. Further issues will emerge and be discussed in the future. For example, the Universal Declaration of Human Rights (proclaimed on December 10, 1948) and the right to access freshwater were encapsulated in the Millennium Development Goals, developed out of the eight chapters of the United Nations Millennium Declaration signed in September 2000. Both have internal and external security implications.[161]

Similarly, factors such as poverty and a lack of opportunity for populations in civil societies have become elements of strategic and security concern. Endemic poverty creates major law and order problems that are often termed "insurgency" by the Establishment, be it in India, Nepal, Bangladesh, or Pakistan. Freshwater availability has contributed (along with other bread and butter issues) by occasionally inflaming the problem. The level of unemployment in the entire subcontinent is set to explode. It is undeniable that unemployment and poverty have contributed in the growth of both terrorism and religious fundamentalism. As people become desperate to find a way out of their miserable existence, the leaders of religious or sociopolitical movements gradually find willing participants among their ranks (such as the success of the Maoists in Nepal).[162]

Indian prime minister Manmohan Singh is on record as saying that the Maoist movement poses the greatest danger to the country. Despite this, the government considers it primarily to be a law and order issue. At the time of writing, the Director General of Police in Chattisgarh (a state in India) was considering whether internationally renowned author and social activist Arundhuti Roy could be prosecuted under the Chattisgarh Special Public Security Act/Unlawful Activities Prevention Act for her essay "Walking with the Comrades," published in *Outlook* magazine earlier in 2010.[163] History tells us that such an approach is always counterproductive, particularly in a democratic country.

Strategic concerns and security issues, therefore, must possess at least three mutually constructive and interdependent elements, namely political-military, economic-social, and environmental aspects. These elements do not affect countries uniformly, but vary significantly depending on the local conditions, even within continental regions such as South Asia.

Looking at the three mutually constructive and interdependent elements, we can see that the environmental aspect is just as important as the other two. The impact of climate change is widespread, and is politically or economically boundary neutral. For example, the destruction of the Amazon forests impacts heavily on regional and global rainfall patterns. Similarly, greenhouse gas emissions contribute to rising sea levels that threaten the existence of low-lying nations, even though their contributions to global emissions may be negligible. The impacts of temperature rise on human health and food production are yet other examples.

While global governance mechanisms are far from ideal in a rapidly changing global economic, political, and climatic environment, the ever-increasing capacity of the UN Security Council (UNSC) to inflict collateral damage or even threaten the very existence of a government that violates standards of the global civil society also influence a country's perception of such concerns. One example is the position adopted by the United States and the European Union with regard to Iran's nuclear program. While many question the effectiveness of the United Nations, they fail to see the direct and indirect psychological influence exercised, particularly by the UNSC, on the strategic concerns and security issues of countries.

The nonavailability of freshwater is possibly the most critical strategic concern globally in the new millennium, and certainly in mainland South Asia. In extreme cases, this concern could lead to a confrontation between riparian countries, because of the possible interruption of the free flow of freshwater to lower riparian countries by deliberate action or other unforeseen factors (such as accidental pollution). One European example was the fire at a Sandoz-owned chemical factory near Basel, Switzerland, on November 1, 1986, as described in chapter 2. In South Asia this remains a major concern, as many river systems in the upper riparian countries are highly polluted.[164]

The first serious strategic concern for the lower riparian countries to China must be to ensure that there is no unilateral interference in the freshwater flow of the perennial rivers. The second strategic concern is to make sure that arrangements are in place to minimize the effects from any accidental pollution into the upper riparian waters, and to provide sufficient warning in the event of any accidental contamination. Finally, the third strategic concern is the protection of existing infrastructures, as any failure to do so could inflict damage on the lower riparian countries in the event of a natural or human-made disaster or terrorist-inflicted damage, such as burst dams, chemical spills, and oil pollution. These are emerging concerns. The *Statesman* newspaper in 2009 reported that the US government received warning from its intelligence agencies that terrorists may target dams and hydropower installations.[165]

In the Asia-Pacific region, the collective interest elements (such as collective security, defense pacts or economic cooperation) have been slow to emerge. James Sperling attributes this to Samuel Huntington's "Clash of Civilizations" and argues that the intersection of five[166] of the world's eight civilizations has generated an antagonistic interpretation of the material balance of power in the region. In other words, while civilizations themselves may not have contributed to generating conflict, their differences may have impeded or prevented such cooperation.

Our understanding of climate and the environment is less than perfect, although it is improving with the expanding study of the science of climatology. For example, the University of Fairbanks in Alaska undertook a study of East Siberian Arctic Shelf sediments and found that methane, trapped in the permafrost is seeping out at the rate of 8 million tons every year due to rising temperatures. The study's coauthor, Professor Natalia Shakhova, argues that any release of just a small fraction of the vast methane held in the ice shelf could trigger an abrupt climate change.[167] Others argue that subset permafrost is losing its ability to be an impermeable cap. However, another scientist from the Max Planck Institute for Evolutionary Anthropology in Leipzig argues that there is no proof that the seepage is increasing. These contradictory opinions show our knowledge about climate science is still in its infancy, but the danger signals are there, and it would be an act of irresponsibility to ignore the forecast consequences.

Concentrations of carbon, methane, nitrous oxide, and halogenated compounds in the stratosphere are the primary cause of the greenhouse phenomenon that has affected climate change with disastrous consequences, including freshwater supplies. In South Asia, the freshwater crisis is primarily due to a rapid rise in population and the need for increased food production. The impacts from climate change on freshwater supply are unclear at this stage, except that they will definitely be detrimental. Notwithstanding a few critics, the climate change phenomenon is real, though some consider it a cyclical phenomenon.[168] Even if this is so, the risk is too high to ignore. The 2008 World Economic Forum in Davos, Switzerland, repeatedly focused on the climate change issue because of its direct impact on freshwater availability, food production, and energy supply.[169]

In the absence of a legal compulsion, dispute resolution primarily depends on mutual trust and understanding. When this is lost, conflicts flare up, even if they are unrelated to water. Egypt has lobbied international organizations not to fund upstream irrigation for fear of losing its own supply.[170] Although water-related disputes rarely trigger conflict directly, all conflict-related crises, however, influence the internal and external policies of the countries involved in the conflict. Once this happens, these policies create externalities, particularly for the neighboring countries which share its riparian water.[171] In such circumstances, conflicts often follow circuitous routes and more often take an ugly form.

India is concerned in its relations with Pakistan about evidence of terrorist infiltration into India, while elements of third country religious fanatics affect its relation with Bangladesh. The BBC reported that the Bangladesh government admitted that terrorist groups might have used its territory to target India.[172]

The irony is that this has been happening in a land where cultures, traditions, and beliefs cut across religious communities, and few people define themselves exclusively through their ancestral faith. The British policy of defining communities along religious lines radically altered these self-perceptions. Winston Churchill hoped to forestall independence by opportunistically supporting Muslim separatism, but instead became instrumental in creating Pakistan.[173]

Although ineptitude and negligence directed British policies in India more than any cynical desire to divide and rule, the British were not above exploiting rivalries. Over the years, many of Britain's decisions stoked Hindu-Muslim tensions, imposing sharp new religious-political identities on Indians.[174] The British subtly used religious diversity as a weapon to create mutual distrust within the community. The Hindu and Muslim leaderships realized this, but, like everywhere else, some people were gullible and stepped into the British trap.

Muhammad Ali Jinnah, far from being an Islamic theocrat, wanted a secular Pakistan.[175] Notwithstanding this secular inclination, it is undeniable that Jinnah positioned himself as the defender of Muslim interest, with Churchill's direct encouragement. The Muslims feared that the secular nationalism of the Indian National Congress was a cover for Hindu dominance.[176] Jinnah failed to create a secular Pakistan, as he died within eighteen months of the country's creation. The ruling oligarchy in Pakistan, namely the landed gentry, the all-powerful army of the newly established state, and the erstwhile powerful bureaucracy found their collective niche in the three A's (Allah, Army, and America). Pankaj Mishra rightly comments that religion was never so deeply and enduringly politicized in India as during the last years of colonial rule. The effects of these days sowed the seeds of the current security concerns of countries in the subcontinent.

The physical division was completed within forty days by British lawyer Cyril Radcliffe. The British concession to Muslim separatism led the Pashtun people to ask for an independent state of Pustus, while the Naga tribes wanted an independent Nagaland, the Sikhs proposed Khalistan, and the Baloch people wanted an independent Baluchistan, beginning the subcontinent's Balkanization process. Predictably, the governments quelled these demands with greater brutality than was perhaps possible under a colonial ruler.[177]

Pakistan formally became an Islamic state in 1956,[178] yet the political Islam found only intermittent support within Pakistan until the 1980s, when the country's military dictator (working with Saudi Arabia and the US Central Intelligence Agency), turned the North-West Frontier province into a base for opposing the Soviet occupation of neighboring Afghanistan.[179] Russia's invasion of Afghanistan provided an excellent opportunity to the West to humiliate the Soviet Union by providing all possible support to the Mujahedeen fighters. The epicenter of this exercise became Pakistan and the border region between Pakistan and Afghanistan. This conflict has unwittingly become a major security concern for both India and Pakistan. The orthodox Muslims used this situation to strengthen their power base to widen their capacity to spread their ideology. To them, India had always been an infidel state. The "hawkish" elements within the Pakistan Army found the situation useful for the strengthening of its power base and for economic gain. But the situation created internal mayhem in Pakistan that today threatens its very existence.

While the West's geopolitical policies during the Cold War are partially responsible for creating fertile ground for the fundamentalist ideas to take root in global Muslim societies, India is not blameless either for these developments within the subcontinent. The ruling elite in India used the Muslim population of some 150 million, at least in the initial years after independence, as their vote bank, but failed to bring them into the modern world, unlike what occurred in Turkey or even in Tunisia.[180] This resulted in many disenchanted Indian Muslims, particularly the youth, placing a tribal loyalty to their faith above democratic secular values, unlike the situation in Europe. This is evident because the terrorist groups, nurtured in Pakistan and Bangladesh, have found it easy to attract local accomplices in India in recent years.

The Indian government's support for a separate civil law regime for the Muslim population, despite the Supreme Court's ruling against this, has contributed to sections of the Indian Muslim population feeling they are different from their Hindu brethren. A major pillar of democratic society is equality before law, irrespective of color, creed, and religion. By allowing Muslims the privilege of being governed by the Islamic Sharia law in civil matters, the government not only failed to bring them into the modern world, but also sowed the seeds of a parallel ghettoized society. It was no wonder they saw themselves as different, and closer to fundamentalists elsewhere who argue for the same privilege.

Terrorists and fundamentalists will use any method to attain their ultimate objective. The September 11, 2001, terrorist attacks on the United States gave a new dimension to the fundamentalist issue in a large section of the world's Muslim population, including the subcontinent. Consequently, a host of local and regional socioeconomic issues lost their separate status, and were subsumed, either directly or indirectly, under the issue of fundamentalism.

Although Osama Bin Laden is credited with the terrorist attacks of September 11, 2001, and with terrorist activities since, personalities such as Hassan al-Banna and Sayyid Qutb of the Muslim Brotherhood in Egypt, and Maulana Maududi, founder of Jamaat-e-Islamia in Pakistan, are equally important forces for radicalizing Islam. Maududi described jihad as one of the central duties of Islam. Even in India, as early as the seventeenth century, some Islamic scholars were preaching radicalism: Ahmad Sirhindi (1564–1624) claimed that Islam and Hinduism were fundamentally incompatible, and one could flourish only at the expense of the other. He was instrumental in reviving orthodox Islamic practices after the death of Mughal emperor Akbar, who had preferred secular policies.[181]

During the initial years of the British Raj (1858–1947), because of pressure from Western scholars and Christian missionaries, the Islamic scholar Mirza Ghulam Ahmad, founder of the Ahmadiyya sect, argued that jihad (as warfare against nonbelievers) had lost its relevance in the modern world, and tarnished the image of Islam. Scholars like Ahmad tried to reconcile Western education and science with the injunctions of the Koran, but the Islamic tradition posed greater difficulties partly because its tenets were more structured and defined compared to Hinduism, which could be interpreted more flexibly.[182]

A recent study argues that the word "da'wa" is basic to the study of Islam. Muslims have used the word throughout the history of Islam and one might translate "invitation to Islam," although the Koran is imprecise on its meaning.[183] The intertwining of this word with the practical world has not been the same over the ages. Islamic missionary movements sought and tried to convert non-Muslims to Islam. This remains a key element of how the Islamic terrorist movement seeks to shape itself in the future. Bangladesh and Pakistan have become victims, partially because of their own doing, as some leadership elements often deliberately used the jihadist elements via the state instrumentalities to attain their political objectives, often as a counter to perceived Indian hegemony.

Unless the Muslim leadership succeeds in synthesizing the basic tenets of Islam with modernity, neither the fanaticism nor the syncretic traditions that have flourished in the subcontinental rural areas could be disentangled from terrorist activity. Once nurtured, religious fanaticism gels easily with the community psyche because fear of the unknown dominates the perceptions of the unsuspecting masses, who can be gullible to the religious rhetoric of the jihadists or to indoctrination.

The issue of human rights has taken center stage in governance policies; any form of religious fanaticism or terrorist activity is unacceptable and, ultimately, self-defeating. Yet, the effect of such activity on economic development is disastrous. Lower levels of economic development in Bangladesh and Pakistan have adversely affected their national security positions.[184]

In August 2005 Islamist militants detonated some five hundred tiny bombs within the space of about an hour in sixty-three of the sixty-four districts in Bangladesh. Such acts, at least temporarily, shift attention from economic development.[185] Economists such as Johann Graf Lambsdorff found that the law and order environment of the country significantly influences investment decisions.[186] Global opinion makers highlight the consequences of poor law and order situations when they occur. For example, after a spate of incidents in Bangladesh in 2003, the *Economist* wrote, "Safe only in the departure lounge—Bangladesh's persecuted religious minorities."[187]

India also has elements of the Maoist movement within its borders. There are about a dozen of secessionist outfits in India's northeast, which is home to more than two hundred ethnic and tribal groups. These conflicts are complex and overlapping.[188] They are dominated by ethnic tension, illegal immigration, and the weight of poverty. Most poverty-stricken people have no safety net for survival, and the income gap between the rich and the poor is increasing. It is impossible to predict the future form and shape of these conflicts.

Along with a deteriorating environment, rising temperature and declining level of water availability, conflict between Kulaks as large landowners and the majority small farmers will certainly intensify, particularly in areas where farmers are suffering most (such as in Telangana in the state of Andhra Pradesh, Vidarbha in the eastern of Maharastra state, parts of Madhya Pradesh, Orissa, northern Bihar, Chattisgarh, and Jhargram subdivision in West Bengal). Suffering caused by the water crisis can take various forms, from petty law and order problems to class struggle. Although not directly tied to the water resource issue, they do have a bearing on how future events inside countries unfold. They are governance and development issues, and water scarcity is an integral part of any governance and development processes.

Assuming that water-sharing issues do not become more complex globally than today, the main security concern will be to protect and secure the dams and reservoirs that store water, both from dam breaches and from polluted water. Particular attention is needed to ensure that, either accidentally or by design, upstream water is not contaminated, making it impossible for lower riparian countries to use the water.

The lower riparian countries must remain vigilant for two reasons. First, there is a long history of violence in all subcontinental countries. Unless the underlying reasons for this violence are addressed, such violence might find its way to assist the activities of would-be terrorists. Second, terrorist activity is taking place in India, Bangladesh, and Pakistan in the name of Islam. It would be unwise to ignore the root causes of such beliefs lest they spread into the freshwater crisis domain.

Two major steps are needed at the intracountry level. First, systems must be put in place to identify and eliminate biological and chemical contaminants and protect inhabitants from exposure. Second, there needs to be in place a combination of physical barriers, real-time biological monitoring and treatment, and the development of smart and integrated response strategies at all levels.[189] Nothing can produce a better result than developing and implementing governance practices that inspire confidence among the masses, because it is they who decide whether terrorists will succeed in their endeavors. Terrorists are very active in India: between January 2004 and March 2007 the death toll from their attacks in India was 3,700, second only to that experienced by Iraq.[190]

In democratic societies often cunning yet legal political practices allow policies that have sinister objectives to enter through the back door. This practice succeeds most where education levels are low, poverty is rampant, and the majority of such population is gullible to simplistic religious propaganda. The Indian subcontinent has most of these. The concept of vote banks is a product of this mechanism. Past Indian voting patterns show that often only 35 percent of the vote ensures electoral victory. Since the Muslims make up upward of 20 percent of the electorate in about 80 (out of 543) parliamentary constituencies, and at least 15 percent in another 40 or so constituencies, such vote banks allow fundamentalists and would-be terrorists to actively use the electoral system to destabilize the very integrity of India's democratic structure.[191]

OPTIONS FOR POLICY INITIATIVES

A concern for each level of government within the subcontinent is to ensure food security, particularly for society's vulnerable groups. Priority areas are increasing productivity, reducing wastage in food processing, producing items that require less water, and adopting new policy initiatives in areas such as nutritional education, and explaining the economic and health reasons for changing food habits. The United Nations has suggested the following three health and nutrition aspects for policy consideration:

- underscore the human face of the food crisis, by monitoring the impact of the food crisis on nutrition, health and poverty, including the effect this may have on reducing or delaying the attainment of the health and nutrition-related Millennium Development Goals, as set by the United Nations;
- provide sound information and analyses for targeting the most vulnerable groups within the most vulnerable countries. There are already countries with high levels of acute and chronic malnutrition, and it is conceivable that the food crisis will hit hardest there; and
- provide health and nutrition inputs for developing or enhancing food aid distribution, as well as social protection activities connected with nutrition for shielding the most vulnerable groups.[192]

No matter how we approach the food security issue, the worst sufferers are the poorer sections of the community.[193] Farmers in developing countries are struggling despite recent rises in the price of the commodities they produce.[194] They will not have the capacity to pay for basic food. Nor will they be able to compete with the rich to cover their food shortfalls. Policy makers must remain aware that the rich, the privileged, and the powerful are always better able to tilt policy directions in their favor. There is virtual unanimity that the inequities suffered by farmers are mostly due to the asymmetries and imbalances of political and economic power.[195]

Famine is not rare to the subcontinent. During the Bengal famine of 1943, the community remained docile and meek. This attitude is no longer the case in today's globalized world, with its changed social and political environment. The reaction would be turbulent, nationally and, possibly, internationally.

Using croplands to produce fuel-producing agents is a complex issue in the subcontinent. Market forces tilt the case in favor of ethanol-producing agents in rich countries. Given that growth in the world cropland base has slowed markedly since the middle of the twentieth century, expanding only one-fifth as fast as the world's population, and given the high purchasing power of car owners, many countries will be tempted to increase the acreage of energy crops, at the expense of food crops. Even mid-tier countries such as Brazil and South Africa are encouraging this development. The subcontinental countries need to be extremely careful, as they have a history of difficult law-and-order situations, even in times of peace. There are many examples of a difficult food situation leading to internal social and political turbulence, and flaring up other issues. Access to basic food at affordable prices is a critical policy issue related to climate change and the water crisis.

Finally, while new policy initiatives are needed for effective outcomes, mass education programs are the most important for two reasons. First, education ensures people understand the complexity of issues that require tough policy decisions. Second, education opens the gate for stakeholder participation in policy development. This creates a congenial environment for introducing changes and implementing new policies. The environmental implications of climate change, on water crises policies in particular, require sustained efforts and ongoing policy initiatives. There is no place for complacency. Complacency means settling for the status quo, and the world can ill afford the luxury of not addressing the problem of climate change.

NOTES

1. Dapice 2009.
2. Durning 1991: 156.
3. Barrett 2008.
4. 2008 World Economic Forum, Davos, Switzerland. The survey interviewed 61,600 people in sixty countries, which included India, Bangladesh, Nepal, and Pakistan in South Asia.
5. http://www.weforum org (accessed January 24, 2007).
6. Issues include the following: protecting the environment, closing the gap between rich and poor countries, war on terrorism, encouraging economic growth and improving the global economy, reducing wars and conflicts, integrating more social issues into the agenda, restoring trust and honesty in government, business and individual, improving/maintaining human rights, overcoming AIDS and other serious health issues, eliminating extreme poverty and hunger in the world, reducing global organized crime, promoting full equality for women around the world, and overcoming the drug problem and drug trafficking. For further details, see http://www.weforum.org/pdf/gallup/vopstats.pdf.
7. Felipe and Miranda 2009: 41–42.
8. See Commission on Human Rights 2003.
9. Estes 1998: 35.
10. United Nations 2005: 234 and 225–27.
11. Srinivasan 2010.

12. Sen 1981: 58–59. The estimate of fatalities referred to is his recent revised estimate; see http://www.indiatogether.org/interviews/sen.htm (accessed January 15, 2009).

13. P. Mishra 2007.

14. Milliman, Broadus, and Gable 1989: 340–45.

15. Sea levels are influenced in part by powerful currents that course around the globe in a pattern, known as *thermohaline circulation.*

16. Yin, Schlesinger, and Stouffer 2009.

17. Schwartz and Randall 2003; King 1983: 119. Also see "Wall St. Underwater: Rising Seas to Hit NY" and "Sea Level Rise due to Global Warming Poses Threat to New York City," http://au.news.yahoo.com/a/-/world/5392287/wall-st-underwater-rising-seas-hit-ny-hard (accessed March 16, 2009) and http:/www.sciencedaily.com/releases/2009/03/090315155112.htm (accessed March 16, 2009).

18. Meehl 2009. Meehl was a draft contributing author in the summary for policy makers (IPCC 2007).

19. "Antarctic Ice Bridge Collapse," http://www.abc.net.au/lateline/content/2008/s2536504.htm (accessed April 7, 2009); "Peter Garrett Joins *Lateline*," http://www.abc.net.au/lateline/content/2008/s2536506.htm (accessed April 7, 2009); "Ice Bridge Ruptures in Antarctic," http://newsvote.bbc.co.uk/mpapps/pagetools/print/news.bbc.co.uk/2/hi/science/nature/7984 (accessed April 7, 2009).

20. Faris 2009.

21. "Maldives President Calls for Reframing Climate Debate," *Times of India*, March 2010.

22. Bhalla 2007.

23. Harrabin 2007.

24. http://news.bbc.co.uk/2/hi/programmes/documentary_archives/6500041.stm (accessed March 29, 2007).

25. Cockburn 2010.

26. Conniff 2007: 28.

27. Hansen 2006a, 2006b.

28. Dupont and Pearman 2006: 50; Gore 2006: 206–7, 58.

29. "An Elephant, Not Tiger (A Special Report on India)," *Economist*, December 13, 2008.

30. World Bank 2002: 134–36, 48–49.

31. Bhutto 1969: 163.

32. Strauss 2002: 143; Nayar 1975: 3.

33. Dr. Malek, political advisor to a former prime minister of Bangladesh (*Northeast Daily*, June 16, 2003), is reported to have said that there was no need for Bangladeshis to migrate to Assam or anywhere else in India, as the living standard in Bangladesh was quite satisfactory! Also see, Hussain 2003: 139; Chapman and Thompson 1995: 187. Occasionally similar view has been expressed by few elected and government officials.

34. CNA 2007: 24; Barlow 2007: 146; editorial in *Asian Age* (October 31, 2008): "Assam Blasts: Was the Govt Sleeping?"; "Politics More Important Than Security," *Times of India*, October 31, 2008.

35. "IB Detects Foreign Hand In," http://www.thestatesman.net/page.news.php?clid=2&theme=&usrsess=1&id=57162 (accessed October 16, 2004).

36. Libiszewski 1999: 134–35.

37. Commission on Human Rights 2003: 21,133, 135.

38. In May 2008, the South African High Court awarded constitutional water rights to the poor.

39. Mahmoud 1999: 47.

40. Faris 2009.

41. Mendelsohn, Dinar, and Sanghi 2006.

42. Pronk 2008: 382–86.

43. Dyson, Cassen, and Visaria 2004: 6.

44. Gijsberts, Hagendoorn, and Scheepers 2003: 242. The study covered twenty-two European countries, including Australia and New Zealand.

45. Lewis 1995: 188.

46. A vote bank is a loyal bloc of votes from a single community who consistently backs a certain candidate or political grouping. Lt. Gen Sinha (Retd.) former governor of Assam, *Statesman Weekly* (November 8, 2008: 2).

47. Dasgupta et al. 2007.

48. United Nations Conference on Trade and Development 1989: 39.

49. United Nations 2007: 65; Giles 2007: 14; and Stern 2009.

50. "Floodwaters Bring Dengue Fever Outbreak," http://www.abc.net.au/7.30/content/2008/s2488980.htm (accessed February 28, 2009).

51. Dr. Eric Chivian shared the 1985 Nobel Peace Prize for spotlighting the effects of nuclear war on global health. He is currently director of Harvard Medical School's Center for Global Health and Environment.

52. "Nobel Laureate Explores Links between Climate Change, Biodiversity": As the Copenhagen climate summit came to an end, Dr. Chivian spoke with PBS *NewsHour* correspondent Paul Solman on December 18, 2009; http://www.pbs.org/newshour/bb/environment/july-dec09/biodiversity_12-18.html (accessed December 29, 2009).

53. Dr. Aaron Bernstein, research associate, Harvard Medical School, and a world authority on the human health implications of global environmental change at an interview with the ABC's (Australia) *Radio National* (*Breakfast Program*) on April 12, 2010.

54. "93% of Wild Medicinal Plants in Endangered List: Study," *Times of India* (April 28, 2010).

55. *Economist* (December 12, 2009). The marine worm is known as sandcastle worm. It secretes glue and uses this to stick bits of sand together to form its casting. The glue does not dissolve in water. It solidifies soon after being secreted. The research is carried out at the University of Utah and headed by Russell Stewart. The second research is conducted by David LaVan of the National Institute of Standards and Technology in Maryland by using the electric eel's ability to generate powerful shocks, which inspired the development of a new type of battery that can run a digital musical player for about ten hours.

56. McMichael et al. 2001.

57. "Global Warming May Lead to More Blue-Green Algae Outbreaks," *Canberra Times* (February 16, 2009: 6). The article is written by E. Kretowicz and contains input from Professor Ian Falconer, an internationally known expert on the subject.

58. Increased travel and trade also often spread many water-borne diseases. A recent report from California confirms that "West Nile virus" was first detected in New York in 1999, most likely through an infected person or bird. Eighty percent of the infected population does not develop symptoms, but a small proportion develops symptoms and becomes victims of severe neurologic diseases. One instance of infection has now spread in another state in five years, and the person has died. This incidence shows the veracity of certain disease that can be sourced to moribund or sterile waters. With intensive agricultural practices in tropical countries, dams are becoming nutrient enriched quickly. The World Health Organization believes that this increases the possibility of excessive aquatic weed growth, including a particular type of microscopic algae. The toxicity of these organisms has been discovered only in recent years. These are potentially lethal to humans and animals if consumed in sufficient quantities (Chorus and Bertram 1999). Other diseases, such as lymphatic filariasis and Japanese encephalitis, are also believed to be rapidly spreading in India and in neighboring countries through weed-infested reservoirs and perennially irrigated rice farming (Hunter et al. 1993: 4–5, 26 and 30–31). Even the much-publicized Indira Gandhi Canal is now considered to be contributing into the increases in stomach ailment for children and an epidemic of skin diseases that was not so prevalent earlier (Blinkhorn and Smith 1995).

59. Participants at the "As Global Temperatures Rise, So Too Do Health Risks" are Dr. D. Campbell-Lendrum and Dr. M. Neira (World Health Organization), Kristie Ebi (IPCC), Dr. J. Patz (director of Global Environmental Health, University of Wisconsin, Madison, and also a lead author of the IPCC), Dr. T. Kjellstrom (Australian Centre for Population Health), and Dr. R. Weber (an allergist at National Jewish Health, Denver). The interview was conducted by Ray Suarez, who covered the Copenhagen summit for PBS *NewsHour*. The program was aired on December 24, 2009.

60. "Worldwide Malaria Situation," http://news.bbc.co.uk/2/hi/health/medical_notes/120644.stm (accessed February 16, 2009).

61. World Health Organization 2008a.

62. Dobson 2009.

63. Postigo 2008: 403–8

64. Patz et al. 2005: 310–17.

65. David et al. 1997: 1341–49.

66. "U.S. Looks to Revive Climate Talks with $100B Yearly Pledge," *NewsHour*, December 17, 2009, http://www.pbs.org/newshour/bb/environment/july-dec09/copenhagen_12-17.html. PBS reporter R. Suarez interviewed D. Campbell-Lendrum, a World Health Organization expert.

67. Hunter et al. 1993: 50–116.

68. Morse and Berger 1992: 325; World Health Organization 2008a, 2008b.

69. World Resources 1998: 26.

70. The information is collated by the author from a group discussion conducted by the Australian Broadcasting Commission in a program known as *Background*

Briefing. The discussion was initially broadcast on August 31, 2008. It was repeated on January 25, 2009, when another victim of such virus attack was identified.

71. "Ministers Gather in Vietnam to Discuss Pandemic Influenza Responses," http://www.abc.net.au/raasiapac/stories/201004/s2878193.htm (accessed April 21, 2010).

72. "Data Belies Health Hope," *Statesman*, February 22, 2007. National family health survey carried out by the Union Health Ministry in India found that "despite the excellent growth record for which India is getting appreciation, its social indicators are dragging mainly due to poor access to health care and infrastructure." Furthermore, "few resources being allocated to the sector and access to health care remains poor." Other regional countries are unlikely to be better.

73. Pacific Institute, "World's Water," 2005, table 2.2, http://www.worldwater.org/table2.2.htm (accessed December 29, 2008).

74. Prüss et al. 2002.

75. Rasmus and Salkowitz 2009: 187–88.

76. "Fixing Health Care: Technology Has Been a Culprit Behind Runaway Health Costs. It Might Now Help Tame Them," *Economist*, April 18, 2009.

77. Agrawal and Srinivasan 2000: 127–66.

78. Tang 2000a: 11–22.

79. http://www.who.int/water_sanitation_health/diseases/burden/en/print.html (accessed December 29, 2008).

80. United Nations Development Programme 2007–2008: 247.

81. United Nations 2005: 219–36.

82. Data for Bhutan were not available.

83. Runge and Senauer 2007: 51.

84. "Health Care in Cuba," http://www.tulane.edu/~rouxbee/kids98/cuba.html (accessed December 15, 2008).

85. "Global Health Council," http://www.globalhealth.org/news/article/4097 (accessed December 15, 2008).

86. Kenneth McLeod, chairman, Department of Bio-engineering, Binghamton University, Binghamton, New York, in the *Economist*, January 24, 2009.

87. Dupont 2000: 46.

88. US Department of Agriculture, Economic Research Service, Food Security Assessment/GFA-14/February 2003: 35.

89. McMullen 2008. The forecast was made by Donald Coaxe, global portfolio strategist at BMO Financial Group, Toronto.

90. Pearson, Blinch, and Kovalyova 2009.

91. Chatterjee 2010.

92. "Draft Food Security Bill May Irk Sonia, States, SC," *Times of India* (March 2, 2010).

93. IRRI 2004.

94. "Nights Getting Warmer in India, Cereal Output May Fall: Study," *Times of India*, April 13, 2010.

95. Kelleher 2008.

96. Chamie 2009.

97. The State of Food and Agriculture: Agriculture Information Bulletin 765/1.

98. United Nations 2005: 234.

99. Vidal 2007a; Ravikumar 2008.

100. Data obtained from the agricultural experts at the Australian Broadcasting Commission's rural reports.

101. "You Are What You Eat," *Economist*, December 19, 2009.

102. Bradsher 2008.

103. US Department of Agriculture, http://www.nationmaster.com/graph_T/agr_gra_ric_pro (accessed February 3, 2009).

104. Brown 2004: 171–81.

105. Speakers at the publicly broadcast interview were Paul Risley, Asia spokesperson, WFP; Professor Beth Woods, chairperson, IRRI, and president of Policy Advisory Council, Australian Council for International Agricultural Research; and Duncan Macintosh, spokesperson, IRRI; http://www.abc.net.au/ra/asiapacific/programs/s2158361.htm (accessed February 9, 2008). Also see Rosenthal, 2008.

106. Bradsher 2008.

107. Dupont 2000: 43.

108. Brown 2004: 185–86.

109. Economic and Social Commission for Asia and the Pacific 1992: 111.

110. Chapman 2000: 109; Chapman and Baker 1992: 30–31, 61–62. Chapman and Baker also cite an engineer familiar with the land degradation via water logging and increased salinity in the Pakistan Punjab such that "heroic measures are essential if the Punjab is not to be destroyed."

111. Mr. S. Aziz in his keynote address to the Pakistan chapter of National Committee of the International Commission on Irrigation and Drainage said that by 1986 about 10 million hectares of land were waterlogged, although the warnings of impending water logging and salinity were evident fifty years earlier, but little was done to implement corrective actions (Jensen, Rangeley, and Dieleman 1986: 32).

112. Brammer 1990a, 1990b.

113. Wade 1985: 465.

114. Rosegrant, Cai, and Cline 2002: 11.

115. McMullen 2008.

116. Barlow and Clark 2002: 60.

117. McMullen 2008.

118. *Economist*, January 3, 2009.

119. Scientific director of the International Programme on the State of the Ocean.

120. Rogers 2010. Also see, "Fin Times: Ban the Trade in Blue Fin Tuna—But Set a Clear Path to Sustainable Exploitation," *Economist*, March 20, 2010.

121. Shearman and Smith 2007: 50–53.

122. Xenopoulos et al. 2005.

123. Ray 2003: 35–38, 166–71.

124. Study findings were presented at the World Economic Forum, Davos, Switzerland, January 2010. The index ranks 163 countries on their performance across twenty-five metrics aggregated into ten categories, including environmental health,

air quality, water resource management, biodiversity and habitat, forestry, fisheries, agriculture, and climate change.

125. Ridgwell and Schmidt 2010.

126. "Universal PDS Only Way to Beat Hunger, Tackle Inflation?" *Times of India*, February 19, 2010.

127. C. Murphy, "Dhaka's Halting Return to Democracy," *Far Eastern Economic Review*, December 2008.

128. Werker 2008: 78–79.

129. *Statesman*, January 8, 2010.

130. Department of Foreign Affairs and Trade 2008.

131. Debroy 2009: 17–20.

132. Australian Bureau of Agricultural and Resource Economics 2007; Darwin 2001; "Climate to Slash Farm Exports Heavily," *Sydney Morning Herald*, December 7, 2007, http://www.smh.com.au (accessed December 13, 2007). Also, see Cline 2007.

133. Pearce 2004.

134. Runge and Senauer 2007: 50.

135. Former British prime minister Tony Blair endorsed it. Ed Gallagher, chairman of Renewable Fuels Agency, endorses it albeit with some face-saving concerns and sops. The British transport secretary Ruth Kelly endorses Gallagher's position.

136. R. Black, "Slowing the Bio-fuels Bonanza: Analysis," BBC, http://newsvote.bbc.co.uk/mpapps/pagetools/print/news.bbc.co.uk/2/hi/science/nature/749 (accessed March 31, 2009).

137. Searchinger et al. 2008.

138. Vidal 2007a.

139. Cline 2007; Kamdar 2007.

140. Revkin 2007.

141. World Health Organization 2008a.

142. Pearce 2007: 40–43.

143. Giggacher 2009: 29.

144. Simons 2007.

145. Chanda 2007b.

146. United Nations 2007: 144.

147. Dr. John McCarthy has carried out research into forest, governance, community-based natural resource management, and agricultural transformation in frontier forest areas of Indonesia for 15 years. Quoted by Giggacher 2009.

148. Howden 2007.

149. F. Seymour, director, Centre for International Forestry Research, at the United Nations Bali climate conference in 2007 (reported in Environment News Service, December 11, 2007). Also see Vidal 2007a.

150. Beresford-Jones 2009.

151. Kugelman and Levenstein 2010.

152. *Economist*, "Touch Wood: Everyone Agrees on the Need to Save Trees, but the Details Are Still Tricky," December 19, 2009.

153. The United Nations Collaborative Programme on Reducing Emissions from Deforestations and Forest Degradation in Developing Countries, http://www.un-redd.org/.

154. "Better Than Nothing" and "Why Farms May Be the New Forests," *Economist*, December 19 and 30, 2009, http://www.economist.com/world/international/displaystory.cfm?story_id=15124802 and http://www.economist.com/world/international/displaystory.cfm?story_id=15179766.

155. Phelps, Webb, and Agarwal 2010.

156. Lorenzoni et al. 2000.

157. Dellink et al. 2009.

158. Programme's Fourth Policy Board Meeting in Nairobi approved $14.7 million on March 22, 2010, according to a press release; http://www.un-redd.org/ (accessed April 23, 2010).

159. Banerjee 1998: 26.

160. Fullilove 2008: viii.

161. The eight Millennium Development Goals are to (1) eradicate extreme poverty and hunger; (2) achieve universal primary education; (3) promote gender equality and empower women; (4) reduce the child mortality rate; (5) improve maternal health; (6) combat HIV/AIDS, malaria, and other diseases; (7) ensure environmental sustainability; and (8) develop a global partnership for development.

162. Lancaster 2003; Haldar 2004.

163. Ghosh 2010.

164. Tributaries of the Hanjiang River (a branch of the Yangtze) in Hubei province or Dongjing River (one of the tributaries of the Yangtze) are examples. A number of media reports by China's own Xinhua News Agency confirms this. The Ganges River system in India is another glaring example.

165. *The Statesman*, March 20, 2009.

166. These are Confucian, Japanese, Hindu, and Islamic and Europe.

167. "Modeling Sub-sea Permafrost in the East Siberian Arctic Shelf: The Dmitry Laptev Strait," *Environmental Research Letters*. doi:10:1088/1748-9326/5/1/o15006.

168. In chapter 7, we refer to Pew Research Center's survey in the United States. The same survey found that 60 percent believed in 2009 that scientific experts still question climate change phenomenon.

169. Rachman 2008.

170. Bate 2006: 3.

171. Rothchild 2002.

172. February 18, 2009.

173. von Tunzelmann 2007: 126–28. For a clearer understanding of the partition legacies, this book is an essential reading.

174. P. Mishra 2007; Chapman 2000: 3–26, 295–96.

175. von Tunzelmann 2007: 80.

176. Ali 1942, 1949.

177. Linter 2010.

178. Bayly 1997.

179. P. Mishra 2007.

180. "India's Radical Islam Problem," *Far Eastern Economic Review*, December 5, 2008.

181. Naqvi 2008: 14; Dhume 2009a: 62–64.

182. Iggers and Wang 2008: 102.

183. Račius 2007: 5, 215–18.

184. Hoffman 1998: 48.

185. "On the Edge of the Knife," *Economist*, December 10, 2005: 67.

186. Ray 2006: 21.

187. It wrote,

> Murder, intimidation and fear of more to come, as 11 Hindus killed in Chittagong; a mosque in Dhaka attended by the Ahmadiya community was attacked by hard-line Sunnis that injured at least 50 people. Members of Bangladesh's mostly Christian or animist indigenous groups have been demonstrating against what they say is the encroachment of Bengali settlers into their ancestral homeland in North-East and South-East. The Awami League claims that these are designed to scarce Hindus into fleeing the country. (November 29, 2003: 128)

188. "The Terror Spreads—India's North-East," *Economist*, October 9, 2004: 107.

189. Gleick 2006: 1–28.

190. Estimate by the National Counterterrorism Center, Washington, DC.

191. "India's Radical Islam Problem," *Far Eastern Economic Review*, December 2008: 7–10.

192. World Health Organization 2008a.

193. A recent worldwide BBC survey unequivocally concluded that the main losers have been poor people who live in cities in poor countries. United Nations (FAO) says thirty-six countries are still in need of external assistance because of continuing local high prices, crop failures, or conflicts; http://newsvote.bbc.co.uk/mpapps/pagetools/print/news.bbc.co.uk/I/hi/business/7340214 (April 17, 2009).

194. Melik 2009.

195. Ingram, Whiteley, and Perry 2008: 17.

Chapter Five

China: A Key Player, an Enigma, or a Pace Setter in the Climate Debate?

China's role in the 2009 Copenhagen Summit was baffling; it gives an impression of a conflicting approach. Chinese president Hu Jintao committed that China should make every endeavor to achieve a win-win outcome at the summit for both developed and developing countries.[1] Later, Chinese Foreign Minister Yang Jiechi claimed that the summit produced significant and positive results.[2] However, some did not hesitate to blame China for the summit's failure. One commentator rightly argued that

> China's decision to resist binding emissions targets at Copenhagen gives a glimpse of a country with big and sometimes conflicting plans for growth, trade and influence.[3] The developed world is aware of the limitations of US policies in the shadows of domestic forces (and, in this case, the response of the Congress and the Senate to the US President Barack Obama's climate bill), but many are unaware of the ongoing battles in the upper echelons of China's government that ultimately shape its policy.[4]

China has set multipurpose goals to achieve its climate change objectives: resources, security, investment opportunities, the strengthening of its support base in its objective of reintegrating Taiwan with the mainland and the return of other areas, achieving recognition as leader of the developing world, and developing export markets. All its moves at the Copenhagen Summit were set within these multipurpose goals. Many China observers informally conceded that China even skillfully used Sudan and Ethiopia during negotiation stages without openly exposing its position.

Until the 1970s China's policies were primarily driven by domestic needs. At that time its strategic goal was to maintain stability in its neighborhood. China realized that it might not be able to solve the difficult issues in the short term, and hence its policy focus was on reconciliatory gestures with the expectation of a positive outcome,[5] while also not hesitating to confirm its firm commitment to issues close to its heart. China has a steadfast and unyielding approach to matters of national interest such as security and sovereignty.

China has gained influence worldwide through carefully crafted strategies in developing nations. It is the preeminent holder of global savings and is in a position to take a leadership role not available to other countries.[6] In April 2010, two important multilateral agencies, the World Bank and the International Monetary Fund, agreed to grant China more influence in the World Bank's loan-making decisions. China promptly welcomed this decision. It provides China with an important platform at an international level to influence many critical development decisions in developing economies.[7]

China remains steadfast on its long-term objectives and, where required, does not hesitate to make subtle shifts to achieve its objectives.[8] China's multidimensional policy initiatives indicate that its approach to international issues, including economic and climate change policies, is inseparable from other policies. China can now make solar panels less expensively than the United States, Europe and Japan, and is believed to be on the verge of achieving parity between solar power and conventionally produced electricity.

Yu Jianrong of the Chinese Academy of Social Sciences argues that social conflict is the taboo of Chinese political discourse. He implies that factors such as social unrest in China, anger in the street, etc., will almost certainly influence direction of future policies.[9]

China is conscious that, in a world where all countries are vying to secure new sources of natural resources, a long-term view of political relationships is essential and that, in the realm of international politics, such changes do not happen overnight.[10] China understands the dynamics of international politics better than most. Its active role in establishing the Shanghai Cooperation Organization (SCO) is one such example.[11] China's decision to fund the development of Gwadar Port is another example, as Baluchistan holds large reserves of oil and gas. It is also of geostrategic importance due to its proximity to Afghanistan and the Caspian region. It is close to the Strait of Hormuz, through which 40 percent of the global oil traffic is routed. However, the United States and Japan question China's motives.[12]

China's policy on Africa is another example. Its trade with Africa increased by 45 percent between 2007 and 2008; and Africa is China's second highest trading partner after the United States. China mainly imports oil and other resources from Africa. With increased trade, it is using its enormous pool of foreign currency to cement diplomatic alliances, secure access to natural resources and source new business. Foreign aid and cut-rate loans with more commercial lines of credit are two of the mechanisms often used by China. Since 2001 China has funded more than two hundred projects in Africa, but it hardly advertises its intentions and focus after developing and executing its policies. One reason for China's easy acceptance in Africa has been its assistance with finance infrastructure projects based on loans with lower interest rates than those on offer by most European nations,[13] which many fail to understand, as the following shows: "We know more about China's military expenditures than we do about its foreign aid. . . . Aid really is a glaring contradiction to the broader trend of China's adherence to international norms. It is so strikingly opaque it really makes one wonder what they are trying to hide."[14]

Politically, China accepts every opportunity to stamp its mark at international forums. It gained immense success in the European Union–Africa summit, held in Lisbon, Portugal, in late 2007, because Europe was unable to quash recriminations from its counterparts over trade, or smother controversy among its own nationals about the invitation to Robert Mugabe, Zimbabwe's president.[15] Collectively, the European Union still cannot agree on how to deal with China.[16]

China's interests in the Sudan are twofold: it has stakes in Sudan's vast oil reserve, and currently Sudan chairs the G77. Further, China has indicated its interest in funding part of a multi-billion-dollar Kenyan project to develop a port and transport corridor that could provide a new export route for oil from Sudan. This achieves two objectives simultaneously: to free south Sudan from its dependency on north Sudan for transporting oil; and to change the dynamics of the long-term peace process in that part of Africa. Such moves confirm that China's policy objectives are to allow constantly evolving policies to meet the challenges of the dynamics of changing international circumstances.

China also wants to lead the world in green technologies. In its western deserts, China is building the world's largest wind farms, the scale of which dwarfs anything in operation in Europe, North America or South America. As the largest producer of solar technology materials, China has been able to reduce the unit cost of solar energy. The 2010 bilateral agreement between India and China on a reforestation program reconfirms China's success in the massive reforestation programs it has completed, and its new reforestation targets to reduce CO_2 gas emissions during the next ten years. By encouraging foreign companies to move factories to its shores, China is achieving the multiple objectives of economic development and climate change mitigation. First, wherever possible, China buys substantial shares in foreign resource companies and second, it restricts the export of certain scarce raw materials that are basic inputs for producing green energy. China has restricted the export of the chemicals dysprosium and terbium, "rare earths" that are vital for a wide range of green energy technologies and military applications such as missiles.[17] China also sees an environmental benefit from this approach, as such shifts will lead to increased production volume and, in turn, to a reduction in China's unit cost of production. It will also create "green energy" products that are competitive with the unit cost of existing energy. By adopting this approach, China will not only achieve its economic development objective, it will also successfully confront the climate change challenge.

As discussed in other parts of this book, internally China has adopted a coordinated approach to minimize CO_2 gas emissions over the years through various policy initiatives. Besides the quality of its environment, China's biggest concern is paucity of water, the supply of which is fixed and unevenly distributed throughout the country. Morally, though not legally, China needs to come to an understanding with other riparian countries over an equitable sharing of riparian waters. This is mostly achievable at the bilateral and regional levels. This approach will certainly test China's political acumen and diplomatic capacity.

As regards the environment, China is the third worst country (after Brazil and the United States).[18] The indicators used are forest loss, habitat conversion, fisheries and other marine captures, fertilizer use, water pollution, carbon emissions from land use, and species threat.[19] China's annual CO_2 per capita gas emission in 2007 was about one-fifth the rate of Australia and the United States, and about one-half of the 2005 European Union rate (see table 1.1). When total emission is considered, China's position was the highest in the world in 2007. China's steadfast refusal to any legally binding agreement is understandable, as it is lagging well behind most industrialized countries and is not prepared to accept any restriction that limits its capacity to achieve rapid economic development, particularly when it is pursuing a proactive gas emission strategy, while also actively promoting alternative energy sources.

CHINA'S FRESHWATER ENVIRONMENT

Tibet is the freshwater source for much of Asia, including China. Since its annexation of Tibet in 1950, China has established its predominance in the freshwater policies of many Asian countries. It has made China the dominant headwater power in Eurasia, giving it control over the upper reaches of Asia's great rivers, some of which flow into South Asia.[20]

By controlling the headwaters, China is in a position to directly influence regional freshwater policy. It is also one of three countries (the other two being Burundi and Turkey) who voted against the draft 1997 UN Convention on the Law of the Non-navigational Uses of International Watercourses. China's decision to oppose the convention is not an isolated incident; it can be seen as a part of China's long-term strategic and economic objectives. China's insistence at the Copenhagen Summit that all figures and numerical targets be stripped out of the accord, even those that did not apply to China, cannot be seen as an isolated position either.[21] By remaining firm on that issue, China ensured that it did not want to be restricted by any international bindings that limited its policy options.

China's policy approach to water management issues, as in other policy areas, is complex, unconventional, and sometimes unpredictable. One thing is certain: all policy approaches by China have been to further long-term objectives.[22] A recent example is China's decision to sign a joint statement to promote trilateral cooperation on water challenges with Japan and Korea in March 2009. These nations stated that, due to climate change and rapid urbanization, they are facing exacerbating water crises (with floods, drought, typhoons, and sea level rise). It is perplexing that China did not become a member of the Mekong River Commission, nor did it address the riparian water sharing issues in any meaningful way with India or Pakistan, two of its nearest neighbors who are facing similar water crises. Instead, it penned the agreement with countries far away from its own regional sphere of influence.[23]

China is ahead of other countries in adopting better water management practices. It has implemented structural changes in water use. In 1949, agriculture consumed 97.1 percent. By 2000, this usage had reduced to 74.1 percent. In industrial and urban areas, consumption rose from 2.3 percent to 20.7 percent and 0.6 percent to 5.2 percent, respectively. Further, although irrigated land increased by one-eighth to 6.7 million hectares, water used for irrigation purposes remained nearly constant (as did water use in the industrial sector—at about 110 bn m^3). Irrigation being the predominant water consumer, improving water productivity became China's critical policy objective. It is also experimenting with new irrigation techniques. In 1997 China introduced a water-saving program on 15.3 million hectares.[24]

China's freshwater demand has been increasing rapidly in line with its rapid economic growth, though China's yearly rise in consumption rate is low compared with other countries at a similar stage of economic development.[25] Per capita 2,241 m^3 of freshwater is available in China, compared to 1,878 m^3 in India. China's annual freshwater withdrawal of 525.5 bm^3 is higher than India's 500 bm^3. China withdraws 18.6 percent of its freshwater compared to India's 26.2 percent. Seventy-seven percent of China's withdrawal is for its agricultural sector compared to India's 92 percent. In urban areas, China has been steadily rising residential water prices.[26]

The estimated average annual runoff of all Chinese rivers is 2,711 bn m^3. Although China ranks sixth in the world after Brazil, the former Soviet

Union, Canada, the United States, and Indonesia, its per capita runoff is only one-quarter of the world mean.[27] Some forecasts indicate that China will have used all its available freshwater by 2030.[28]

China owns 48 percent of the Brahmaputra basin and a little more than 11 percent of the Indus river basin (see table 2.2).[29] Both India and China have made further claims on the basin. If the Mekong experience is any guide, China will use the waters of the Brahmaputra at a time of its choosing.[30] Currently, India and Pakistan share the waters of the Indus, and both are trying to harvest additional water from the system. Given China's political closeness with Pakistan, in any conflict between India and Pakistan on the water-sharing arrangements its sympathy will probably lie with Pakistan.[31]

China cannot meet all its agricultural water needs in its arid and semiarid regions. In a number of those regions scarcity has reached a critical point. The effects from drought are noticeable in the inhabited areas, and the trend of drought intensifies as human activities increase. According to T. G. Rawski, a China expert, the severity of such droughts depends on macroeconomic adjustments (especially employment growth).[32] Unsustainable water withdrawal from rivers is affecting the water environment and ecosystems downstream. In coastal areas this has led to seawater intrusion.

In southern China, with its humid climate, population pressure has led to an increase in the number of rice crops (from one to two) which, in turn, has resulted in frequent drought. In the north China plain, with an annual precipitation rate of between 600 mm and 800 mm with highly developed agriculture, drought impacts have been critical since the 1970s, as small and medium rivers run dry. The surplus of groundwater has resulted in environmental degradation, and the frequency of drought continues to rise.[33]

Since 1985 the Yellow River has failed to reach the sea for part of almost every year. Sometimes it does not even reach the last province on its journey to the sea. A few small rivers no longer exist, and almost 1,000 lakes have disappeared in Hebei Province alone. The groundwater level is falling by 2–3 m a year in the north China plains. China's northern wheat belt has suffered from droughts for decades, caused by pervasive overuse and waste. Aquifers have depleted and, in some regions, wells probe a half-mile down before striking water.[34] Beijing is also facing a severe water crisis where groundwater levels are falling by two meters each year.[35]

Chinese scientists have warned that if the rate of temperature rise does not decline, the glaciers on the Qinghai-Tibet Plateau will shrink by 2030, accelerating the freshwater crisis in China.[36]

China has adopted a basinwide approach to data collection and to the planning and management of the water of its main rivers. Looking at China's water management practices long term, the institutional patterns established by the nationalist government provided a base, despite transformations and upheavals in China during the latter half of the twentieth century.[37] One BBC journalist argued notwithstanding toxic power stations, burgeoning a number of cars, and a determination to achieve economic growth, China is creating real, long-term environmental solutions.[38]

China has achieved significant growth in water productivity: a mere 1 percent increase in water use achieved 9 percent economic growth. A feature of China's water policy is that it is increasingly becoming proactive. It has established a water policy framework that since the 1980s has included two important legislative texts stipulating standards for governing resettlement and other related issues:

> The water law adopted in 1988 aims at adjusting the state's policy towards water activities: utilization of water, control against disasters, harnessing of rivers etc.
>
> Regulations passed in 1991 [saw] compensation for land administration, [and] resettlement for the construction of water conservation and hydroelectric projects. On land administration, consultation with affected communities is required.[39]

However, in the areas of regulatory and operational functions, China's achievements remain wanting.[40] For example, in only six of China's twenty-seven largest cities does the quality of drinking water meet government standards, and the groundwater in twenty-three of those cities fails to meet the standards.[41]

RIPARIAN FRESHWATER POLICY

The freshwater policies of riparian countries are reflected in the direction of their respective foreign policies, economic interests, and security concerns. Examples are China's role in the reconstituted Mekong River Commission, and Egypt's efforts in ensuring that it continues to receive the major share of water from the Nile, even at the expense of upper riparian counties.

With the post-Mao economic reforms, China changed and a new paradigm emerged. With China's rising status as an economic and political powerhouse, it started believing in economic strength as a basis for both military power and diplomatic influence in the new world order.[42] It has increasingly become assertive in its external and economic policies.[43] For most of the past two decades, its approach to international issues has been clear, as stated by the late Deng Xiaoping: "China would like to be number one, but it would still rather get there without making enemies: keep a low profile; not take the lead, watch development patiently and keep your capabilities hidden."[44]

China has settled its land border disputes with all its neighbors except India. Japan and Vietnam have complained about what they allege to be Chinese encroachments in disputed islands and atolls in the South China Sea and in the Pacific Ocean. Both China and Japan are claiming the ownership of the Chunxiao gas field in the East China Sea, as Japan sees it as lying only 4 km from the border of its Exclusive Economic Zone. The reason for conflict appears to be the availability of natural gas and oil in the disputed area. China is also in dispute with Japan with respect to the status of an atoll in the Pacific Ocean. China's dispute with Vietnam relates to the ownership of Paracel Islands, an archipelago in the South China Sea. Taiwan has also lodged a claim for this archipelago, although not as forcefully as Vietnam.

Many dynasties have ruled China, beginning with the Hsia dynasty and ending with the Manchu dynasty. The Mongols ruled China before 1912 when the Nationalists took control, ending in a revolutionary government in 1949. Many of China's calls for border readjustment could be the product of decisions taken by earlier rulers that China is now contesting. If a similar principle is followed elsewhere, then many current independent nations will cease to exist, particularly in Asia, Africa, and the Middle East.

China has surpassed many countries in military spending, both within and outside the Asian region. The balance of influence between China and the United States is shifting in favor of China. China has been driving a wedge in Taiwan between the business community and the administration by pursuing three links: trade, transport, and communication.[45]

Since the 1962 border skirmish, bilateral relations between China and India have improved, but not blossomed, nor have contentious issues been resolved. In recent negotiations, China often has adopted an inconsistent approach.[46] For example, China refused to issue visas to two Indian government delegates from India's Arunachal Pradesh, allegedly on the grounds that it is disputed territory. In 2009 China upped the ante by voting against a US$2.9 bn Asian Development Bank (ADB) loan to India, a small portion of which was to be used for irrigation projects in Arunachal Pradesh. India initially mustered enough support within the ADB to override Chinese objections, but, in subsequent negotiations, with support from Japan, Korea, and Australia, China effectively blocked the portion of the loan meant for the disputed province.[47] China's undeclared rivalry with India for economic and political leadership in the region may have been responsible for this outcome.[48]

Trade interests often act as a buffer in difficult situations between countries. Trade between China and India has been negligible in the past. Earlier it was little more than 10 percent of the value of the trade between China and the Association of Southeast Asian Nations (ASEAN).[49] In recent years trade has been flourishing. The increase of US$14.2 billion during the first five months of 2007 was, at the time, the fastest growth rate achieved by China among its top ten trading partners.[50]

In the economic area China earned regional respect for its active efforts to ensure that the regional countries were not disadvantaged by China's economic response to the Asian economic crisis of the 1990s. China's dealings with Thailand, Myanmar and Cambodia are examples.[51] This became clear during the 2008–2009 global economic crises, when China had ample opportunities to play a leading role, but quietly ensured the safety of its trillion dollar reserves.[52] Domestically and regionally, the economic damage was kept to a minimum. It is important to understand that "commerce affects the domestic politics in the smaller state, shaping politics and definitions of national interest. Firms and sectors form political coalitions, create vested interest in target states, and reshape governmental policies in different ways than if they had been affected by domestic pressures alone."[53]

China's economic power is reflected in its newly acquired position as the biggest exporter in the world.[54] The gradual decline of the European Union in global affairs and the relative economic decline of the United States have offered China opportunities to play a more proactive role in international affairs, both at regional and global spheres.[55] China asserts that it does not interfere with the internal affairs of other countries, but often appears to do the exact opposite.[56] The Indian media recently quoted a Chinese government website that reflects a different policy approach. The website states, "[So] it is baseless to say that we will not set up any military bases in future because we have never sent troops abroad. . . . It is our right. . . . As for the military aspect, we shall be able to conduct the retaliatory attack within the country or at the neighborhood area of our potential enemies."[57]

The Pakistan-Afghanistan conflict and the rise of Taliban in its neighborhood may have prompted its move away from its earlier stated position, but Chinese ambiguity and an inconsistent approach in a number of international matters raise concern in countries where bilateral relations have been less than friendly.

CHINA AND THE REGION: RIPARIAN WATER-SHARING COMPLEXITIES

India has not hidden its concern over China's increasing economic and political role, particularly in Nepal and Pakistan. The role of these two countries in India's freshwater policy is critically important. Traditionally Nepal was considered a close ally of India. China, through its "boutique" policy mixes, has been able to reduce Nepal's dependence on India in a number of areas.

China has firmly established its "friendly credentials" in the hearts and minds of large sections of the Nepalese community. India has a 1950 friendship treaty and a number of water resource–related treaties with Nepal. But treaties can be rescinded, and China can easily play a spoiler's role from behind the screen.

Nepal was exclusively dependent on India for transit rights to the outside world. On occasions India interfered with this right, which caused strong resentment against India. China is keen on minimizing Nepal's dependency on India for transit rights, and has been providing infrastructure assistance to achieve this.

In 2009, the Indian media reported that China intends to build a dry port east of Kathmandu, with a parking space for two hundred trucks and a *go-down* (a dockside warehouse in India and East Asia) to place goods after customs clearance.[58] Nepal's former Maoist prime minister, Prachanda, is reported to have asked China to extend its rail connection from Lhasa in Tibet, the birthplace of Buddha, to the dry port.[59] Should it eventuate, Nepal will not have to exclusively depend on India for transit rights; a new trade route might encourage regional economic development and provide an opportunity for China to be more intimately involved with Nepalese development policies.

All Nepalese rivers feed the Ganges, which is vital to India's economy. The 1990 Nepal law on using Nepal's natural resources required approval by a two-thirds vote in parliament, and was directed, many believe, at restricting India's access to Nepal's water projects. This limits India's capacity to enter into bilateral arrangements with Nepal for developing its water resources for mutual benefit.

Following the fall of the monarchy in Nepal and the parliamentary election in August 2008, the Unified Communist Party of Nepal (Maoist) became the largest party in Nepal's parliament. Although Maoists are no longer in government, Prachanda as the party's leader has railed against the Indian government on numerous occasions.[60] An increasing Maoist insurgency within India and the alleged political links between the Maoists in Nepal and India remain a major concern for India. Publicly, India has stated that it falls to the people of Nepal on who they elect in their parliament, yet a perception remains that India has been able to influence Nepal's policies by indirectly supporting political parties such as the Nepalese Congress (which, incidentally, was established in Kolkata in 1949). Perceptions play an important role in shaping political attitudes everywhere.[61]

India-Pakistan relations have never blossomed since 1947, and both nations have remained antagonistic to each other ever since. Further, after the border skirmish of 1962, China's foreign policy tilted toward Pakistan. Pakistan has found a strong political ally, and China has rewarded this loyalty by siding with Pakistan on many contentious issues, against India's expressed wishes.

Pakistan is a strong Islamic country and is a lower riparian country to India. It has looked to China for moral and political support at crucial junctures in its history.[62] Managing the China-Pakistan axis on further accessing the Indus water for development purpose is a challenging strategic concern for India.

India's national security concern has increased since the rise of the Taliban in Afghanistan and Pakistan's consistent support to terrorists. China has not taken sides on insurgency and terrorist issues, so India remains unsure about China's position in any preemptive action against terrorists. The Afghan conflict, the rise of the Taliban, and the role of the Inter-services Intelligence in Pakistan's administration does not contribute to the genuine promotion of détente between these two countries. A. J. Tellis, a senior associate at the Carnegie Endowment for International Peace, writes, "Flushed with confidence flowing from the success of the anti-Soviet jihad in Afghanistan during the 1980s, Pakistan sought to replicate in the east what it had managed in the west. . . . The Pakistan military has no interest in dismantling any terrorist assets that it believes serve it well."[63]

The Pakistan-Taliban issue will not disappear for two reasons. First, the ISI provides strong material and moral support to the Taliban. Even if the North Atlantic Treaty Organization (NATO) and the United States are able to destroy the Taliban externally, it will be difficult to do so within Pakistan. Second, it is widely believed that the Taliban leadership has decentralized its activities, including devolving its leadership structure. This makes it difficult to separate the Taliban elements from the population at large. Further, Pakistan is facing a serious socioeconomic problem; a recent survey of young adults found a very disturbing and unsettling environment.[64] To divert its internal sociopolitical tensions to an outside element that binds all against a common enemy suits Pakistan well. A push toward Indian hegemony against rogue states would encourage Pakistan to strengthen its bond with China.

Pakistan and China stand by each other. China's political acumen may encourage it to cultivate Pakistan as a trusted ally. It is in China's political interest to keep India engaged with Pakistan. India could use its bond with Pakistan to get engaged in the thickening webs of trade, investment, and technology transfer, and to achieve its dual objective of becoming a regional political and global economic power.[65] China's long history indicates that it rarely takes its eye off the main goal, but can simultaneously pursue an inconsistent policy: "Chinese history has consisted of a series of changes in national self-image. . . . The use of force, self-strengthening, the isolation policy, the signing of treaties, and the domestic mass movements were all elements of the cybernetic regulation, which has served to actualize and confirm the national self-image."[66]

China proceeded with the construction of the Three Gorges Dam on the Yangtze River despite strong global opposition.[67] It displaced some eleven million people, directly affected the lives of twenty million, and destroyed 62,000 acres of farmland, thirteen major cities, 140 large farms, and countless villages.[68] China ignored the findings of the independent evaluation report commissioned by the dam opponents that challenged the official evaluation report supporting the dam.[69] China has managed the situation cleverly by ensuring that industrialized countries participate in the financially lucrative construction contracts.[70]

In its Nu River project in western China, it took a very different position. The project, designed to harness hydropower through a thirteen-stage dam, would have destroyed World Heritage–listed areas of unique biodiversity and displaced an estimated fifty thousand people, mostly ethnic minorities.[71] Although China ignored all earlier opposition on such issues, a plausible explanation has been that it wanted a prestige-enhancing Green Olympics in Beijing in 2008, and did not want to confront the minority tribal groups who strongly opposed the project.[72]

Another of China's approaches is to keep the views of experts within a narrow private domain and not permit public debate, as a politicized solution would not fit with its monolithic state apparatus.[73] One needs to be aware of these issues to understand China's approach to riparian waters.

The freshwater crisis has intensified in Pakistan. Recently, Sardar Aseff Ali, the Deputy Chairman of Pakistan's Planning Commission until January 13, 2008, and an adviser to Pakistan prime minister Syed Yousuf Raza Gilani, declared: "[India] will have to stop stealing Pakistan's water as the latter will not hesitate to wage war."[74]

It is worth noting that Pakistan lost an arbitration case recently under the dispute resolution procedures of the 1960 Indus arrangement. The statement is troubling, because in the past governments in Pakistan have tried to divert peoples' attention to external issues to hide their government's own policy failures.

India, Bhutan and Bangladesh share the Brahmaputra basin. The Brahmaputra has surplus water.[75] Bangladesh and India need the Brahmaputra waters to bolster water flow in the lower Ganges below the Farakka Barrage, but cannot agree on the most appropriate mechanism to do so.[76] Any unilateral decision by India to use the surplus water may encourage China to make decisions unilaterally on how it could extract maximum benefits from the Brahmaputra waters. The Brahmaputra water issue cannot be resolved without China's involvement. China's recent announcement confirming its initiatives to use Brahmaputra's water and related policy issues are intriguing.

For some time the Indian media has been reporting that China was building one or more dams on the Brahmaputra River for irrigation and generating hydropower. China has routinely denied this. Even as recently as 2009, China denied that it was constructing the Zangmu dam in Tibet's Lhoka prefecture.[77] Recently China admitted that it was building a hydropower project on the river. Ostensibly, this change in attitude is prompted by the unprecedented cooperation of India and China during the Copenhagen Summit in 2009.

The Zangmu dam is designed to produce 510 MW of electricity, will not involve storage of water and was a run-of-the-river project. Informed sources confirmed that there would be four more such projects, but all within Chinese territory.

As noted earlier, China voted against the draft 1997 UN Convention on the Law of the Non-navigational Uses of International Watercourses. In general, this understanding on sharing riparian water is that an upper riparian country can use the water without unduly harming the interests of lower riparian countries.[78] A hidden element in this principle is that a river is a living entity and its flow should not be interrupted without fully examining the consequences of any such interruption. A matter of concern is that apparently China had made it clear that it did not really have to share its plans with India, but that it was doing it out of a sense of trust. With this pronouncement, China was denying the implied principle in any riparian water-sharing arrangement. China and India will now share the hydrological data on the Sutlej and Brahmaputra rivers, although agreements were signed in 2005 and 2008, respectfully.

China's assertion that it did not share the information earlier because it was building the dam within Chinese territory could carry a hidden meaning. Does it then imply that if China diverts the waters of the Brahmaputra to feed its parched northeast in the future, which is possible with modern technology, the same argument will persist? In not becoming a member of the Mekong River Commission, China is clearly indicating that its long-term national interest takes precedence over all other issues. Quizzed on this issue in the parliament in 2009, the government of India clarified its view on China's long-term plans by replying that China had clarified that it was a responsible country and would never do anything to undermine any other country's interest. Mutual trust is an important element in building confidence between parties. In this context, China's assurance is welcome, but India needs to remember that it is also an upper riparian country and is building a similar project on the Sutlej, an action against which Pakistan has been threatening terrorist action.[79] Nevertheless, every trust-building initiative concerning water-sharing arrangements is welcome.

FLEXIBILITY, AN OVERPOWERING KEY ELEMENT IN CHINA'S POLICY APPROACH

In December 1978, then Chinese leader Deng Xiaoping suggested that modernizing agriculture, industry, national defense, and science and technology are essential (mantras) for reaching number one position in the world. The modern leadership added one more to these four, namely "flexibility," designed to make the leadership's task easy. China has used this most successfully when dealing with complex issues to achieve their national objectives, but has been rigid on its long-term interests.[80] One example is the way China used the US embargo on arms sales to Pakistan following the Indo-Pakistani war of 1971 to reverse US policy on China.[81]

In other instances, such as the problems in western China where many people live in water-short, degraded, and overpopulated conditions, China has adopted a flexible region-specific policy approach, but has ensured the patronage element of the policy does not become prominent.[82] A similar approach characterized China's much-maligned population policy, which permitted significant ethnic, spatial, and temporal variations in places such as Tibet and Xinjiang.[83] In its drive to achieve rapid economic growth, China has not hesitated to marginalize farmers—the backbone of the Chinese revolution—despite their protests.[84]

Realistically, one has to agree with Akihiko Tanaka's observation that it is impossible to judge in which direction China will proceed.[85] However, one can argue that because of its central control, and assuming this can be sustained, China can (with its large surplus of funds) enlist the labor of millions to build major infrastructure and other projects, possibly for decades to come. This makes China relentlessly dynamic in ways that other democracies, with all their temporizing, are not. This dynamism will drive China's policies abroad, because of its need to secure energy and strategic minerals to support the rising living standards of its immense population.[86]

The policies of international countries toward China should inevitably be nuanced in nature, focusing on either China's potential domination or internal chaos. To the extent that there is hope of China becoming interdependent with the rest of the world and a more open society, the policies of neighboring countries should encourage such positive moves. If such interdependence were to eventuate, riparian water-sharing arrangements would prove an easy task.

Unfortunately, China's demonstrated capacity to ignore the spirit of international conventions leads many to believe that current international water laws may be unable to handle the ongoing problems for riparian countries.[87] The major strategic concern for all riparian countries in the region, therefore, is how best to reconcile their respective country's interests with those of China. This is possibly the most complex of the tasks facing large and small countries. However, as China will remain a monolithic state, the situation is unlikely to change in the near future.[88]

The Indian Foreign Ministry reported during Premier Wen Jiabao's visit in mid-April 2005 that he had agreed to improve bilateral relations, including water resource management. In response, one commentator argued that, just as India bore the brunt of the rise of international terror due to its geographical location, so it would be directly affected by the growing power of an opaque, calculating empire next door.[89] In this context, the press statement by Zhou Gang, an advisor to the Chinese Foreign Ministry and a former ambassador to India, is interesting:[90]

> India must make [a] "substantial adjustment" in the eastern sector . . . as the Chinese people will never accept [the] "McMahon Line." . . . [During my term as Chinese Ambassador in India] I made it clear on many occasions [that] Tawang belongs to China, it is the birthplace of the 6th Dalai Lama and the Dalai Lama is China's Dalai Lama, who cannot be India's Dalai Lama. . . . If the Indian side can make substantial [adjustments] in the East sector according to the principles of mutual understanding, mutual accommodation and mutual adjustment, the Chinese side will make its adjustment in the western sector accordingly.

In view of these confusing relationships, a foreign policy focus on riparian water-sharing issue has to be a strategic priority for India. Strategic concerns and security issues will certainly be different for both China and India. One needs to keep in mind that national policies alone may be of little help to the vulnerable groups who are living under fragile economic and health conditions.[91]

China's focus on achieving economic growth and protecting its security needs may encourage it to adopt a more inclusive approach to water-sharing and border settlement issues.[92] Yet, China may consider it politically advantageous to keep some contentious issues alive. In either event, sharing the waters of the Brahmaputra becomes a major strategic concern for both India and Bangladesh.

NOTES

1. Goldenberg 2009.
2. http://news.yahoo.com/s/ap/20091220/ap_on_sc/climate (accessed December 21, 2009).
3. Browne 2010.

4. Jianrong Yu (2007), director of the Institute of Rural Development at Chinese Academy of Social Sciences, has written extensively on this issue. Some of these are social unrest in China, anger in the street, farmers having the right to keep land for themselves, large-scale farmers' protests sweeping the countryside, and social conflict in rural China.

5. Gancheng 2005: 71–75.

6. James 2009: 167–68.

7. "China Welcomes Increased Influence at World Bank," http://www.abc.net.au/ra/asiapac/stories/201004/s2883099.htm (accessed April 27, 2010).

8. See "China's Thing about Numbers," *Economist*, http://www.economist.com/world/international/displaystory.cfm?story_id=15179774 (accessed January 23, 2010).

9. Yu 2007.

10. Margolis 2006; Caspary 2008; Dingle 2009.

11. Members of the Shanghai Cooperation Organization are Republic of Kazakhstan, People's Republic of China, Kyrgyz Republic, Russian Federation, Republic of Tajikistan, and Republic of Uzbekistan. The entire region is considered to be very rich with enormous oil reserves, and many consider that China is securing its own resource position in the post–Cold War and post-Taliban periods (Vines 2007: 213–19; Bakshi 2004: 284–85, 303).

12. "CHINA: Leaked US Report Reveals Scale of China's Military Ambition," http://www.abc.net.au/ra/asiapac/programs/s1285151.htm (accessed January 25, 2005).

13. Sautman and Hairong 2010.

14. LaFraniere and Grobler 2009.

15. Wallis 2008; Orlik and Yong 2009: 56; "How China Sees the World," *Economist* (March 21, 2009; April 18, 2009).

16. "Charlemagne I Dragon Nightmares," *Economist*, April 18, 2009.

17. Oster 2009; Grammaticas 2009; and Bradsher 2009.

18. Bradshaw, Giam, and Sodhi 2010. Also see Adelaide University 2010.

19. Bradshaw, Giam, and Sodhi 2010.

20. Richardson 2008.

21. "Banyan: From Charm to the Offensive," *Economist*, January 9, 2010.

22. Chey 2007.

23. http://www.chinaview.cn (accessed March 20, 2009).

24. Zhang, Ke, and Zhang 2005: 336.

25. Gleick 2003: 191.

26. "Bottling It: Water Pricing in China," *Economist*, January 9, 2010.

27. Zhang, Ke, and Zhang 2005: 322.

28. United Nations 2007: 97; Buckley 2007.

29. Barlow 2007: 146.

30. Yardley 2007.

31. See "China Goes into Overdrive," *Statesman Weekly*, April 7, 2007, for China's reaction following its admission as an observer in the SAARC. "Dam Disagree-

ment Affected Talks," *Statesman*, January 28, 2005. At a meeting with editors of the Asia News Network in Davos: "Dam Disagreement Affected Talks."

32. Professor T. G. Rawski, at a seminar November 18, 2009, in the Department of Political and Social Change at the Australian National University, Canberra. The title of his seminar was "China and the Environment—Tempest in a Teapot." Interested readers may wish to consult the following two books on China: *China's Rise and the Balance of Influence in Asia* (Keller and Rawski 2007) and *China's Great Economic Transformation* (Brandt and Rawski 2008).

33. Zhang et al. 2005: 324–25.

34. Wines 2009.

35. Houghton 2007: 160.

36. Buckley 2007.

37. Pietz 2006: 92–119.

38. Jonathan Porritt, December 2, 2009, from his dispatch from Copenhagen climate change meeting. Porritt was unequivocal that China was effectively leap-frogging the older industrial societies of Europe and America and bringing on real long-term solutions. He further suggested that amid the toxic power stations and burgeoning number of cars, he found some extraordinary and pioneering green solutions.

39. Padovani 2006: 120–43.

40. Frederiksen, Berkoff, and Barber 1993: xviii.

41. United Nations Economic and Social Council 2004.

42. Hoffman 1998: 48.

43. Buzan 2004: 164; Bernstein and Munro 1997; and Shambaugh 1996.

44. "A Time for Muscle-Flexing," *Economist*, March 21, 2009.

45. Keller and Rawski 2007a: 7, 150–51; and Segal 2007.

46. Chellaney 2004, 2005.

47. Dhume 2009b.

48. Tanham 1992: vii.

49. Chung 2004: 117–18, 170.

50. Statistics released by the general administration of customs in Beijing on June 11, 2007. Reported in the *Statesman* on the following day.

51. Cody 2005.

52. Godement 2009; Schell 2009; Jiang 2009.

53. Mastanduno 2003.

54. "China Becomes Biggest Exporter, Edging Out Germany," http://news.yahoo.com/s/ap/20100110/ap_on_bi_ge/as_china_trade (accessed January 11, 2010).

55. Also see "Charlemagne: Europe and Inscrutable China," *Economist*, January 23, 2010.

56. Ching 2009.

57. Dasgupta 2010.

58. "China Mulls Building Dry Port Near Kathmandu," *Statesman*, January 12, 2009.

59. *Statesman*, March 19, 2009.

60. "Powerplay: Prachanda Slams India," *Statesman*, January 7, 2010.

61. The *Economist* indirectly argued in similar lines when it published a report with the heading "Back to the Brink: The Spread of Violence in Nepal Is Not Just the Maoist Fault," December 12, 2009.

62. *Yale Global*, July 24, 2004. Also, Shih 1990: 123.

63. Tellis 2010.

64. "Front Line against the Taliban," *Economist*, November 28, 2009.

65. Keller and Rawski 2007a.

66. Shih 1990: 123.

67. White 2004.

68. Jhaveri 1988: 56–63.

69. Barber and Ryder 1993: 22–32.

70. Probe International undertook this independent assessment. Pearce (1992), Rich (1994), McCully (1996), and others confirm the enormous power of the dam industry–related professional bodies to preserve their self-interest.

71. Yardley 2004a, 2004b.

72. The entire river basin is inhabited by a tribal Muslim population whose economic life is supported by this river system, and it was opposed to this project. China had brushed aside many such objections, but this time it must have considered the global political environment in post–September 11 and post-Iraq situations and concluded that the risk of alienating such a large group of population at this point in time, given the country's long-term economic and political interests and the 2008 Olympics, is not worth taking.

73. Gready 2004: 9.

74. "Now, Pak Threatens Water War," *Statesman*, January 4, 2010.

75. Houghton 2007: 157.

76. Ray 1998.

77. "China Admits to Brahmaputra Project," *Times of India*, April 23, 2010.

78. Ray 2008a.

79. Ray 2008a.

80. Its approaches to Tibet's autonomy issue or the Taiwan issue are examples.

81. Chapman and Baker 1992: 70.

82. Naughton 2004: 253–95.

83. Huang and Yang 2004: 193–225.

84. *New York Times*, December 8 and 31, 2004.

85. Tanaka 1995: 58–59.

86. Kaplan 2010: 23–24.

87. Boyle 1996: 48; Dellapenna 1999; Gleick 2000: 213.

88. Glaser 2007; Sorman 2007.

89. Chellaney 2005.

90. As reported by the Press Trust of India from Beijing on October 31, 2007.

91. Sachs 2008: 332–37; Someshawr 2008: 366–74.

92. Ferdinand 2002: 130–45; Yahuda 2002: 102–13.

Chapter Six

Regional Perspectives

We argue in this chapter that the Copenhagen Summit did not provide the expected outcome. Most participating countries and nongovernment organizations have given their reasons for this failure. Individually, each participating agency was selective in apportioning blames to others when it suited them. Rarely was there a genuine attempt to find the root cause for the failure, although each partially identified some valid reasons. Sweden's Tallberg Foundation comes very close to the root cause of the failure, although some may disagree. It argues that nature does not recognize nations and, by implication the climate change issue is a geophysical problem that requires a fundamentally different approach.[1]

The foundation's basic argument that everything is interconnected is true, but at the same time it is equally true that nations' political boundaries are a reality and each nation usually tries to protect its own interests during any negotiation. For example, although the creation of the European Union placed the individual rights of member nations below EU rights, the Greek financial meltdown exposed the reality that some nations consider issues of national interest first, before considering them in the EU context.

The subcontinental South Asian Association for Regional Cooperation (SAARC) is riddled with mistrust and political intrigue. We have alluded to the factors behind this mistrust in chapters 3 and 6. If the SAARC member countries cannot forsake their veil of mistrust, it will be difficult to address climate change and allied policy issues in any meaningful way. India had argued at the 2010 SAARC summit in Bhutan that India would hold the post-Copenhagen environment dialogue in a meaningful manner in accordance with the principles of the United Nations Framework on Climate Change (UNFCC) and the Bali Action Plan.[2] This position existed in Copenhagen, but even the summit failed. What guarantee is there that it will succeed now? A measure of innovative thinking is required.

The internal sociopolitical-economic environment of a country influences its approach to climate change and related issues such as riparian water-sharing arrangements. In the following sections we examine each member country's socioeconomic-political environment. In principle this gives an indication of whether a meaningful coordinated approach by all countries can be adopted at the Cancun meeting in late 2010.

Climate change, including the freshwater crisis, is a key security and foreign policy issue for all SAARC countries. This situation is further complicated by the fact that, collectively, SAARC countries have the dubious distinction of having the largest number of people living below the global poverty line. Climate change aside, even in the best of times, drought, flood, poverty, poor governance, corruption, and political conflicts are common phenomena in these countries.

Of the eight member countries, Sri Lanka and the Maldives are physically located outside the mainland. Rising sea levels are threatening the physical existence of the entire Maldives and parts of Sri Lanka. Sri Lanka does not have any riparian river. As the Maldives is increasingly threatened by the rising sea level, people will increasingly face involuntary displacement. There is no regional policy initiative from SAARC, except a recent diplomatic initiative by the Maldives government as outlined in President Mohamed Rasheed's address to the inaugural session of the sixteenth SAARC Summit.[3] In any involuntary displacement, the initial burden falls on the adjoining countries. India has heartbreaking painful experience of accepting thousands of refugees from Eastern Pakistan following independence in 1947. A regional approach will certainly provide additional weight to the initiative adopted by the Maldives.

The situation is less critical in Sri Lanka. In the event of involuntary displacements, some displaced persons certainly can be relocated inside the country. However, its decades-long political strife may prompt some displaced sections to leave Sri Lanka. As most Tamils have a religious and ethnic affinity with Tamils in India, there exists the probability of them targeting India for resettlement. In Tamil Nadu, at least one political party has publicly urged the Indian government to accept Tamil refugees.

Bhutan, the smallest of the subcontinental countries, became a constitutional monarchy in 2008. Its problems are miniscule, compared to other countries in the subcontinent. Moreover, the political steps taken by the ruling monarch has made the country more stable and responsive to the wishes of the people. Bhutan has environmental problems, but every indication is there that it will be willing to constructively contribute to the collective decision of the SAARC countries.

Afghanistan, the newest SAARC member, is heavily preoccupied with its internal problems, and, at this stage, has scarcely a voice, and therefore minimal effect in SAARC affairs. India and Pakistan have never been on good terms. Some in Bangladesh and Nepal believe that India is a hegemon far too ready to throw its weight around. Sections of their communities, including the government, have continuous grievances about India's response to various political situations. Following the abolition of the monarchy, Nepal became a republic in May 2008 and the Maoists won the election. Nepal remains in political turmoil, despite Maoists being outside the government. The current Interim Constitution of Nepal came into effect on January 15, 2007. A new constitution is being drafted, but whether it will satisfy all constituents remains to be seen. After many years of instability, Bangladesh appears to have achieved some degree of stability since the election of the present government.

Despite India's active armed support to Bangladesh's independence struggle, some sections of the Bangladesh community remain antagonistic toward India and consider India a hegemonic country. Two major consequences of India's support were, first, China felt that it had lost to India and redoubled its efforts to intensify political penetration in the region and second, the United States started basing its Asian policy directly on China. However, in the new millennium, subtle shifts in US policy have taken place in the belief that China is becoming too powerful for the United States.[4] The recent press report that India and the United States met to discuss Beijing's modernization of its 2.25 million strong armed forces and its strategic moves in the Asia-Pacific region is one example.[5]

Bangladesh's struggle for independence occurred at a time when the two largest countries in the subcontinent—India and Pakistan—were in two Cold War camps: the Soviet bloc and the West, respectively. Within this context, many concluded that it was the beginning of the disintegration of Pakistan.[6] If anything, initially it strengthened the democratic process in Pakistan, although eventually the army clawed its way back into power.

The internal political situations in Pakistan and Bangladesh have been dominated by the army since their birth, with an occasional elected civilian government. Both these countries have now elected governments, but political stability in Pakistan is far from assured in the shadows of the Taliban infiltration in Pakistan. Both Bangladesh and Pakistan are important to India for environmental management and the resolution of freshwater-sharing issues.

The Cold War has waned, but a new menace of religious fundamentalism has raised its head in all these countries, as elsewhere in the world. In South Asia, it is very visible in Pakistan and it has spread its tentacles deep inside the country. The rise of the Mujahideen in Afghanistan to fight the Soviet

Union with active support and assistance from Pakistan and the United States led to their spectacular rise under the Taliban canopy. From Afghanistan they infiltrated Pakistan and, it is alleged that with active assistance and support from the Pakistan government and its ISI they infiltrated India.

The yet unresolved Kashmir issue gave the infiltrators fertile ground for ready acceptance by certain sections of the community. This has further changed the political environment and chemistry of the relationship between India and Pakistan. There is strong Western pressure on both India and Pakistan to settle contentious issues between them through negotiations, but it is most improbable that Pakistan will agree to proceed without the Taliban in its ranks, while India will not accept the Taliban's presence in any dialogue between India and Pakistan.[7] Climate change issues and water policy are caught in this melee.

As a freshwater shortage becomes critical for both countries, they have started working on harvesting additional waters from the Indus river systems. Thus, water directly entered the political equation of these two countries and, as on earlier occasions, it is now shrouded among other issues. When India decided to construct the Baglihar Dam in the southern Doda district of Jammu and Kashmir, on one of the tributaries of the Indus, Pakistan objected. In line with the 1960 Indus Waters Treaty, the issue was arbitrated and, with modifications India was allowed to construct the dam. This did not placate some sections of the Pakistan Administration and the community, and they now take every opportunity to mix the issues.

China's spectacular economic and political rise has brought a new dimension to the riparian water-sharing issue. In chapter 5 we discuss the implications of China's economic and political ascendency and its impact on the policies of subcontinental countries. China's ascendency has wider ramifications for two reasons. First, China has its own political and economic ambitions. Second, it is predicted that within a decade or so, China's economic and military strength will be competitive with the United States. It will then create its own dynamics in intra- and interregional relations. Within this context, how China will respond to its environmental and water needs becomes an important policy consideration.

POOR REGIONAL GOODWILL IN THE SUBCONTINENT

Contrary to Asian countries, the European countries are more region-focused, assisted by their institutional innovations.[8] The European Bank for Reconstruction and Development (EBRD), the North Atlantic Treaty Organization (NATO) and the European Union contributed to the development of a regional and cooperative approach to address existing and emerging climatic, water sharing, social, economic, and political issues. The European Economic Community established the European Environment Agency in 1994, with a comprehensive brief to cover air pollution, biodiversity, climate change, land use and water. In 2006 sixteen European organizations established the European Water Partnership, an independent nonprofit organization, to find solutions for the water challenges in wider Europe. These are all interrelated issues and one cannot be meaningfully addressed without considering the other issues simultaneously.

Europeans mostly believe that the modern security agenda is rooted in two mutually constitutive and interdependent elements—the political-military and the economic. Some voluntary transfer of sovereignty facilitates the solution of joint complex issues. In South Asia, such understandings do not exist, and the underlying international political complexities and intrigues have worsened the situation. It is a political failure that, despite the existence of SAARC for nearly three decades, complex issues such as environment management and water-sharing issues are yet to be addressed collectively. The leadership cannot blame their failures entirely on the international power play, as many political leaders have found it convenient to do so.

The root cause of continued tensions in South Asia is divergent security perspectives.[9] Their strategic concerns and security issues are not—and can never be the same. Besides harvesting water from common sources, illegal border crossings, harboring and encouraging fundamentalist infiltration, the Kashmir issue, and the transit right are other principal concerns. There are numerous reasons for this:

- wide variation in physical size, population, and economy (such variations influence the psyche of small countries' leadership). The smaller countries often, and sometimes without justification, feel insecure about big neighbors, irrespective of ground realities and, on occasion, seek to use the situation to their own advantage;
- artificial boundaries (the absence of natural boundaries), which contribute to boundary intrusions and skirmishes, spoiling any chance of good neighborly relations;
- identity crises created by a mixture of nationalism, religious conservatism, feudal social structure, and mutual distrust of neighbors;
- religious fundamentalism that creates indistinguishable gray areas, mixing religious values, governance issues and societal values; and
- emergence of superpowers in the neighborhood that opens opportunities for policy manipulation by third countries.

Agriculture dominates South Asian economies, the success of which is key to their economic growth. Agricultural productivity becomes an important element in their economic prosperity. Issues such as environmental degradation, climate change, and freshwater availability are key elements in their endeavor to achieve this prosperity. There are also other requirements, such as land reform;[10] availability of credit, access to markets, and reasonable prices for produce. These issues are outside the purview of this book. This chapter examines the context of their respective socioeconomic-political environment to understand the reasons for their approach to climate change and regional water-sharing issues.

Although Asia's major rivers originate in China, the country is also a water-scarce country. This gives China legitimacy to harvest some of these waters within the precincts of the legitimate interests of lower riparian countries. In the absence of an international legal framework, it is difficult to assess how much weight will be given to this convention by economically and politically powerful countries.[11] In this context it is worth noting that China voted against the 1997 UN Convention on the Law of Non-navigational Uses of the International Watercourse.

The Chinese leadership remains committed to pursuing extensive economic growth policies.[12] The need to produce more food and industrial raw material puts pressure on freshwater harvests. There are five layers in China's political system: national, provincial, municipal, county, and township. The growth in Gross Domestic Product in these spheres is the formal yardstick of measuring success, but, informally, local officials' self-interests play a part, as they personally benefit financially from the growth by investing in key firms.[13] Consequently, even if the central government consents to abide by any internationally agreed decision that might negatively impinge on the interests of the minor layers, experience shows that local officials flout such agreements.

Further, China is unlikely to adopt political pluralism at this stage.[14] Its political system makes it convenient for it to deal with riparian countries without any internal debate and in a secretive way. In its extreme form the emphasis is on national interest as the only criterion, making it possible for China to ignore other issues such as human rights, cultural heritage, and international obligations. However, at the same time, China may consciously adopt a position that the legal reforms of the modern human rights movement often deliver only empty parchment promises to the poor,[15] and as such, this issue can be easily discarded. Such a stance raises issues of political economy as an important element in contemporary climate change, and environmental and riparian water-sharing issues.

Countries adopt foreign policy decisions with multidimensional objectives. Climate change and other related issues need to have more permanency than foreign policy objectives. This can only be achieved with active cooperation between the riparian countries. Unfortunately, such cooperation has been missing from the policy agenda of most subcontinental countries, since each country is mostly overshadowed first, by mutual distrust of each other and, second, by international political power plays.

In the following sections of the chapter, we examine the internal socioeconomic and political environment of each mainland country.

AFGHANISTAN

Afghanistan is a semidesert country with an area of 652,090 km^2 and unique physiographic and climatic characteristics. Its northwestern, western and southern border areas are primarily desert plains and rocky ranges. Its southeast and northeast borders rise progressively higher into the major glacier-covered peaks of the Hindu Kush mountain ranges. Deserts receive less than 100 mm of rain a year, while the mountainous regions receive more than 1,000 mm, mostly in the form of snow. Temperatures vary between seasons and even within a single day. In the northern valleys, summer temperatures rise as high as 49°C, while midwinter temperatures in the Hindu Kush region can drop as low as 9°C.

Afghanistan has twenty-one rivers, and four major river systems and most terminate within the country, except for the Kabul River, which joins the Indus in Pakistan. Afghanistan does not share riparian water with any other South Asian nation.

Iran, Russia and Pakistan are its closest neighbors. These multicountry borders have made Afghanistan a strategic country that has contributed to the current situation on the ground. The residents of the Pakistan-Afghan border are mostly Pashtuns, a fiercely independent tribe. During the creation of Pakistan, Pashtuns demanded a separate state. Their relations with Pakistan have been most difficult since the defeat of the Soviet Union at the hands of the Mujahideen.

Afghanistan and Pakistan belong to the Sunni branch of Islam, but they belong to different subsects. Afghanis belong to the Hanbali and Maliki subsects, while most Pakistanis belong to the Shafi'i sect. These divisions remained dormant when the Mujahideen were fighting the infidel Soviet Union, heavily assisted by the West and the Pakistan government (particularly, the powerful Inter-services Intelligence [ISI]) unit staffed by officers from all three military branches.

The Mujahideen elements mostly consisted of a large number of fiercely independent Afghan tribal groupings. Leaders of each of these groups are warlords, who have hardly any concept of modern governance issues, human rights, and the fundamentals of a modern economy. The economic base of their individual conclaves is poppy cultivation and trading in illegal drugs. Islam, Muhammad and Allah were the boundaries of their cultural emancipation and limits of judicial jurisprudence. With the retreat of the Soviet Union, their common enemy—traditional rivalry between warlords—resurfaced, and the historical rivalry between Afghanistan and Pakistan flared up again.

Internal Afghan rivalry turned into a civil war between warlords. The coalition between warring groups was not permanent; these were shifting coalitions, based on loyalties that would ensure military and political survival, economic gain and the attainment of religious objectives. These culminated in the triumph of the Taliban, a loose confederation of Islamic fundamentalists whose philosophical base was governance based on the Sharia.

Afghanistan started receiving Soviet aid in 1973. The king, Mohammed Zahir Shah, was deposed in 1973, to be replaced by a Russian-backed government. In the closing days of 1979, the Soviet Union directly intervened, by invading the country and establishing a government of its own choice.[16] The stage was set for a brutal power struggle between the Cold War antagonists. The Soviet Union became entangled in Afghanistan on December 24, 1979. An all-out assault by the Mujahideen saw the Soviet forces withdraw on February 15, 1989, just before the end of the Cold War. This action contributed to the rise of the Taliban and eventually led to the current role of outside forces in Afghanistan.

Afghanistan is still recovering from the devastating Soviet invasion followed by Taliban rule. A combined NATO force of Western democracies dethroned the Taliban-led government and installed an elected government in the capital Kabul. The government is now reeling under a regrouping of Taliban and Al Qaeda forces.[17] However, remnants of Taliban still maintain pockets of resistance and wield political power. The situation remains fluid and the failure of the Pakistan government to confront the fundamentalist and Taliban training grounds along the Pakistan-Afghanistan border area has seen the Taliban move farther into Pakistan.[18] A combined NATO and US-led force is now fighting to defeat the Taliban insurgents.

The Afghanistan government and the Western alliance have claimed that the insurgents received training and material support from the ISI, the intelligence unit within the Pakistan Army. They further claim that the Taliban maintains a strong base in Pakistan, from whence they plan operations in southern Afghanistan; although Pakistan routinely refutes such.[19] During the Soviet invasion Pakistan used such elements to fight the Soviets, and now they are in opposition. Al Qaeda and Taliban terrorists have established firm roots in the inhospitable Afghanistan-Pakistan border region, in parts of Afghanistan and also inside Pakistan.

Afghanistan remains extremely unstable. Its strategic location and the interplay of complex international political maneuvering have created significant security concerns for other countries, but particularly for India, as the Taliban try to organize and infiltrate India through Pakistan. Indirectly these maneuvers impact on freshwater-sharing issues in the riparian countries, particularly in any further harvesting of the Indus waters between India and Pakistan.

BANGLADESH

Bangladesh came into existence in 1971 with India's military assistance, initially as a secular state riding the wave of Bengali nationalism. General Muhammad Zia-al-Huq assumed the presidency in July 1977 and adopted Islam as the state religion, thus discarding the basic principle of the founding leaders of the country.

Bangladesh shares fifty-four rivers with India. Almost all its rivers originate outside the country; only 9 percent of its total freshwater resources originate within the country. It has three major river systems, the Ganges (local name *Padma*), the Brahmaputra (local name *Jamuna*), and the Meghna. Other major rivers are the Dhaleswari, the Surma, the Kusiyara, the Karnafuli, and the Teesta. Bangladesh has a 4,000 km–long porous border with India.

Bangladesh has plenty of aquifers, but the water is contaminated by arsenic (a naturally occurring poison)—something that even the World Bank did not realize until it became a serious health issue. Arsenic has already penetrated the food chain.

Bangladesh's average rainfall varies from a maximum of 5,690 mm in the northeast to a minimum of 1,110 mm in the west. It suffers from an abundance of water and regular floods in the rainy season, but also from scarcity of water in the dry season. Floods, cyclones, and droughts are common occurrences.

The effects of climate change are disastrous for the country, and possibly beyond its economic and technical capabilities to overcome the crisis on its own. A large section of the country is low-lying and will be drowned by the predicted rising sea levels. Already some islands in the Sundarbans area in the Gangetic delta (the largest single block of tidal halophytic mangrove forest in the world), and in southern parts of the country have disappeared. Its impact on biodiversity is disastrous, impacting heavily on the habitat of the rapidly declining numbers of the Royal Bengal tiger. A recent World Wildlife Fund report forecasts that sea level would rise 11.2 in. (above 2000 levels) by 2070. This would result in shrinkage of the Bangladesh Sundarban by 96 percent within half a century, reducing the tiger population there to 20.[20]

In 2000, Bangladesh had a population of 131 million. Its density is one of the highest in the world. By 2050, the population will rise to 212.5 million. Assuming no land will be lost due to climate change (which is unlikely), the density per km^2 of land in 2050 will be 1,496.5 persons, compared to 1,007 persons in 2000. Rural density per km^2 of arable land in Bangladesh in 2005 reached 1,259 persons.[21] Its per capita arable land of 0.1 hectare in 1981 had decreased to 0.065 hectares by 2005.[22] The Bangladesh government believes that the booming population is a major obstacle to overcoming poverty.[23] It is one of the world's six countries that account for half the world's annual increase of 77 million people.[24]

Eighty percent of its population live on US$2 or below a day, and nearly 49 percent live on US$1.25.[25] The richest 20 percent receive about 43 percent of the total income, while the poorest 20 percent receive about 9 percent. Corruption and poor governance remain entrenched within the Bangladesh Administration, irrespective of the type and character of the government.[26] Bangladesh has faced severe law-and-order problems. A former president, Badruddoza Chowdhury, has even warned that the country might end up as a failed state, unless the government tackles crime, poverty, and corruption.[27] His Excellency Mr. Anwar Chowdhury, the British Envoy to Bangladesh also made a statement that the high number of illegal activities and graft were damaging Bangladesh's image.[28] Transparency International, a global nongovernment organization based in Berlin ranked Bangladesh the world's most corrupt country for a number of years.

Agriculture is the backbone of its economy. This sector employed just over half of Bangladesh's labor force in 2005.[29] During the last quarter of the twentieth century, Bangladesh developed a garment industry that became the largest foreign exchange earner and a major employer of semiskilled women. The industry lost its special export quota in 2005 when the global garment-industry market was reorganized.[30]

Since the assassination of Bangladesh's founder Sheikh Mujibur Rahaman on August 15, 1975, the Bangladesh Army has dominated Bangladesh politics. The army has ruled Bangladesh sporadically for about fifteen years since 1975. Freedom fighters largely formed the basis of the Awami League and it broadly subscribes to progressive views on national and international issues (albeit in a limited way). Zia established the BNP and its political base has been the conservative elements of the Bangladesh Army that remain ideologically and culturally close to the ideals of the Pakistan Army. It has also retained the largely conservative, anti-Indian orientation of the Pakistan Army.

The surviving daughter of the Awami League's founder and the wife of the BNP's founder assumed leadership of these two parties. Both parties have maintained their bitter animosity, and each blames the other for all national failures. The rivalry has debased Bangladesh politics to such an extent that the *Economist* commented, "The two parties had debased the country's politics into a Punch-and-Judy show of noncooperation and vindictive retaliation."[31]

The 2007 army coup leaders charged the leaders of both parties with corruption during their occupation of Treasury. The general public appeared to be content with the possible exile of these dynastic leaderships, as the treatment meted out to them was seen as retribution for the corruption, abuse of power, and constant bickering.[32] Political commentators believed this would destroy the two political dynasties and allow the rise of other political forces.[33] However, some considered the entire episode to be like chopping the branches of a tree while keeping the roots intact.[34] The 2008 election returned the Awami League to the Treasury benches. The BNP's share of 300 seats fell from 193 to around 27, while Jamaat-e-Islamia, BNP's largest Islamist alliance partner, won only 2 seats compared to 17 in the outgoing parliament.

For democracy to flourish beyond the mere formal level, it is essential to have free and unfettered associability, genuine social and political pluralism, and commitment to civil society values. In this context this election outcome is positive. Some argue that Bangladesh's political history and traditions were not congenial to the growth of a true pluralist-democratic system because of the prevailing socioeconomic conditions.[35] While Bangladesh did not become a basket case in the mid-1970s, as predicted by the then US Secretary of State Henry Kissinger, some, nevertheless, consider Bangladesh a failed state.[36] It is too early to predict whether the new government can improve the political environment and allow democratic institutions to flourish.

The Farakka Barrage issue has remained a cause for conflict between Bangladesh and India. The problem has existed since 1850, but the loss of the Chittagong port to Pakistan in 1947 made it urgent for India to ensure that the navigation channel to Kolkata Port remained open. If sufficient waters do not flow through the lower Ganges River during the dry season, the port will virtually cease to exist, and the economies of the entire eastern India, Bhutan, and Nepal will suffer inordinately.

From the 1850s until independence, the government undertook eight separate investigations to assess the state of the Ganges River in lower Bengal. All but one committee concluded that augmentation of water was required in the lower Ganges to save Kolkata Port. Dredging the channel would not produce the desired outcome without the feeder channel that supplies water to flush out the silt. The dredging commenced in 1906.[37]

India and Bangladesh signed the Farakka Water Sharing Treaty on December 12, 1996. They agreed that there was insufficient water flowing during the dry months to meet current needs. They need to find a way to increase the flow, but both countries have adopted different positions, particularly with respect to diverting the surplus Brahmaputra waters to the Ganges River.[38]

The treaty stipulates that should the flow in the Farakka Barrage fall below fifty thousand cusecs in any ten-day period during the lean season, India and Bangladesh will enter into immediate consultations. Bangladesh accused India of violating the treaty when the flow remained below the required level for a few days, and wanted India to assure a flow of thirty-five thousand cusecs irrespective of the actual flow.[39] Even after confirming at one international forum the need for improved regional political relations in resolving water-sharing arrangements, Bangladesh raised similar issues at another international forum immediately afterward, fully aware of India's opposition to raising bilateral issues in a multilateral forum.[40]

Bangladesh has often used the Farakka issue to bolster its domestic political positions by raising the alleged Indian hegemony to hide its own policy failures (e.g., President Hussain Muhammad Ershad, a former military ruler, used it during the devastating 1988 flood). When public opinion turned against him, he did not hesitate to describe India as the main culprit, refusing to accept any assistance from India and making a fresh attempt to internationalize the water-sharing issue with India.[41] Rehman Sobhan, a well-known Bangladeshi development economist argued that internal political developments play a major role in the freshwater policies of each of the riparian countries in South Asia.[42]

It is undeniable that Bangladesh's economic security and prosperity largely depends on riparian water flows. The Farakka Barrage's effect on the dry season flow is real, and so is the problem for India. For Bangladesh to blame India for all its own ills and internal policy failures is counterproductive.[43] The political uncertainty does not contribute to addressing water-sharing issues constructively.[44]

Bangladesh, like all other countries, has its own internal security issues. Common triggers of conflicts in most developing countries include

- economic and technological underdevelopment;
- inequality;
- corrupt and undemocratic political structures which result in insecurity and alienation; and
- covert intervention by neighboring states or by others.[45]

What makes countries in conflict particularly different from other countries is the frustration and feeling of hopelessness amongst their largely poverty-stricken people. Factors that can threaten security include

- scarcity of resources;
- poor governance;
- claims of exclusivity in religion and ideology; and
- selective preference to cultural constituencies.

It is now established that a section of the Bangladesh establishment and a number of religious fanatics have close links to fundamentalist elements. Some even believe that the ISI is cultivating links with those elements to work on its behalf in India and that India has arrested a few such infiltrators.

The discovery by the Bangladesh police of an Islamic theological school established by a British charity hiding a cache of weapons is a revelation. The police found firearms, ammunition, and explosive manufacturing provisions in teachers' quarters and in the hidden ceilings of the rooms (including books on jihad). Local officials agreed that foreigners were crossing the border, but nobody knew their purpose for frequent border crossings.[46]

This has created a feeling of insecurity in the region and beyond.[47] To make the situation worse, problems have arisen from the rebellion by the Bangladesh Rifles (a paramilitary border security force) and its confrontation with the army leadership in the cantonment immediately after the elected government took office. A number of army officers and their family members died at the hands of the Bangladesh Rifles forces. These incidents indicate that there exists a major security problem in the country and is bound to spill over into India or put adverse color on bilateral relations.

Another sensitive security issue is the illegal border crossing. Primarily for economic reasons, Bangladesh's large unskilled labor often crosses illegally into India,[48] although Bangladesh routinely denies this.[49] The high population growth rate in some border districts in India during 1991–2001 confirms this infiltration.[50]

Illegal crossing through the long porous border are difficult to manage, and impossible to make a distinction between the populations of these two countries. The country was divided in 1947 without giving due consideration to issues that contribute to the growth of a successful nation: "East Pakistan was created by dividing Bengal in a way which made no sense on any grounds other than to give self-determination to the two religious communities—but even then the new boundary did not succeed as well as it could have done if that was the priority aim."[51]

After partition, Pakistan's leaders, including Bangladesh's founder Sheikh Mujibur Rahaman, started campaigning for extra land from India to accommodate Pakistan's burgeoning population in the east.[52] Such campaigns have fed mistrust at the community level in India. India alone cannot redress this issue, as the United Nations found while examining illegal African migration in Europe.[53]

The terrorist attacks of September 11, 2001, have added a new dimension to this issue. Since the United States mounted pressure on Pakistan to curb terrorist activities on its soil, there has been a resurgence of fundamentalist and militant activities in Bangladesh. Initially the BNP-led government was slow or unwilling to curb the militant's activities. With the increased vigilante killings, international pressure forced the government to clamp down on its activities. This led many to argue that Bangladesh is the next battlefield for political and militant Islam.[54] However, given the socioeconomic environment, it is arguable whether the government would be able to clamp down on the activities of the fundamentalists, without first attacking the root cause. Christine Fair, a South Asia and terrorism expert at the US Institute of Peace argues that even an army might find it difficult to stop such activities.[55]

The central focus of Bangladesh's security concerns is political stability, strengthening the pillars of its governance structure, and ensuring rapid economic development. Both the climate change and water scarcity issues need to be addressed as a priority. This requires a farsightedness and imagination in negotiations, regionally and bilaterally. This will not be easy, as relations between India and Bangladesh have not followed a predictable pattern over the years. Reasons for this include history, geographical location, religious groupings, lack of governance, poverty, and international politics. Political commentators have argued that both governments must act quickly to improve ties with neighboring countries and address these long-standing problems that affect development, as the foreign aid is drying up for poorer nations.[56]

For India and Bangladesh, climate change and water remain strategic concerns. The prognosis does not appear to be bright, as very few writers in Bangladesh are able to refer to the Farakka Barrage in an objective way.[57] The governments must be aware that climate change will only exacerbate issues, triggering economic disruption and create social tension and political antagonism on an unprecedented scale.[58] Bangladesh's constitutional identity as an Islamic country may have contributed to developing this attitude, particularly when religion is used in an oppressive way, Sulak Sivaraksa argued, "It identifies with certain groups, cliques, and certain elements in the name of nationalism, in the name of development, and in the name of whatever it is."[59]

BHUTAN

Bhutan, a landlocked country, is situated in India's north in the Himalayan foothills. Tibet ruled Bhutan from the sixteenth century until the Chinese took over in 1720.[60] The British East India Company concluded a treaty with Bhutan in 1774 and the first hereditary monarchy came into existence in 1907. On India's independence, Bhutan signed a treaty with India in which it agreed to seek India's advice on foreign relations, but not to remain bound to accept it.

Most Bhutan rivers originate in the Himalayas and drain into the Brahmaputra after crossing into India. The four main rivers of Bhutan are the Torsa (*Amo Chu* in Bhutanese), the Raidak (*Wang Chu*), the Sankosh (*Mo Chu*), and the Manas (*Mangde Chu*).[61]

Bhutan's economic base is agriculture and forestry with a limited number of tourists visiting each year. The UN's human development index ranked Bhutan higher than Pakistan, Bangladesh, and Nepal in 2005.[62] Bhutan has fewer poverty-stricken people compared to other countries of South Asia; during 2000–2006 only 26.2 percent of the population were living on US$1.25 a day or less, while 49.5 percent were living on US$2 a day or less.[63]

The kingdom has little social or political complexity. Currently a number of Bhutanese refugees living in Nepal want to return to Bhutan, but India is refusing to grant them transit permits. It alleges that these refugees are ethnically Nepalese. The United Nations is trying to relocate them to third countries. One media report indicates that the United States has agreed to rehabilitate some sixty thousand refugees.

The greenhouse effect has adversely affected Bhutan's freshwater supply and has contributed to a loss of topsoil. The snow on Himalayan glaciers in Bhutan is receding and consequently the rivers originating there receive less freshwater during the nonrainy season. The intensity of monsoon rain is increasing and becoming more irregular. With the rise in temperature, the evaporation rate is increasing with adverse impacts on land productivity.

The parched and desiccated soil is causing a significant loss of topsoil. This is a critical issue, because preservation of topsoil is critically important for an agriculture-dominated economy seeking to preserve land productivity. The washed-away topsoil creates a further problem, as loose topsoil flows into the Brahmaputra River creating regular floods in northeast India and Bangladesh, and causing loss of property, and life.

The monarch has a good understanding of the political psyche of the people in his country and in the region. The king's abdication of the throne in favor of his son, and the establishment of a constitutional monarchy, ushered a new era in Bhutan's relations with India.

Immediately after accession to the throne, the new king indicated his intentions to update the Indo-Bhutanese Friendship Treaty of 1949. The Indian government views this treaty as an historic moment in India's relations with Bhutan.[64] However, given the realities of international politics, and given the underlying rivalry between India and China, there is no certainty that the situation will stay the same in the future.[65]

NEPAL

Nepal, sitting between the two most populous countries on earth, is a landlocked country in the Himalayan foothills. To Nepal's north is China; on the other three sides it is bordered by India. Nepal's geographical location puts it in a strategic position. A special bilateral relationship has existed between Nepal and India that goes back to the days of Britain and the Indo-Nepal Treaty of Peace and Friendship of 1950, including sensitive agreements relating to riparian waters. The free border movement and trade between these two countries have cultivated relations similar to that of a sovereign country and semi protectorate.[66]

In 1769 Nepal became a kingdom following the emergence of the Gurkhas as a force uniting the small principalities. It became a British-dependent buffer state in 1816 and fully independent in 1923. Between 1846 and 1951, hereditary prime ministers of the Rana family ruled Nepal. Following the overthrow of the Ranas, the monarchy returned to Nepal in 1951. Autocratic rule and exploitation of the poor have been features of Nepal's history. This situation improved after Bir Bikram Sha Dev's accession to the throne in 1972 following the death of his father, but discontent remained. Nepal proclaimed a constitutional monarchy in 1990 and the first multiparty election took place in 1991, but political instability continued. The major political parties became infamous for infighting and corruption, resulting in twelve governments in thirteen years.

The police and judiciary remained beyond the common persons' reach.[67] Entrenched poverty and the failure of political parties as mediating structures between the state and society have been blamed for social discontent.[68] The political instability intensified after King's death. With the Maoist insurgency, the country plunged into anarchy. The number of casualties on both sides mounted, extrajudicial killings became the norm, economic development stalled, and social unrest remained unabated.

The Maoist rebels have come from poverty-stricken rural areas where revolution has provided hope for a better life. Despite the Maoists' strong-arm tactics in raising funds for the movement, the population did not oppose it, because of social inequalities and the perception that the government in Kathmandu is not truly representative of the less-privileged castes and the hill people.[69] Rebels have given the movement a strong philosophical underpinning by claiming inspiration from former Chinese leader Mao Zedong, and by tapping into opposition to India's economic and cultural dominance in Nepal.

Only 17 percent of Nepal's hilly terrain is suitable for cultivation. Per capita agricultural land halved between 1981 and 2005, from 0.2 hectares to 0.1 hectares. About 79 percent of the population is engaged in agriculture, an economic environment that guarantees the poverty of the masses. Nepal has an established tourist industry, but its environmental sustainability in the climate change context is uncertain. It is one of the poorest countries on earth. During 2000–2006, 55.1 percent of Nepal's population lived on US$1.25 or less a day, while 49.5 percent lived on US$2 or less a day.[70]

About 15 percent of the total groundwater runoff from Nepal's rivers comes from Tibet.[71] Almost all Nepalese rivers flow to or through India to the sea. Nepal's rivers feed the Ganges River, the lifeline of India's northern and eastern heartland.[72] All Nepalese rivers are tributaries of the Ganges River system.[73] The Ganges River is also important to Bangladesh: hence it remains strongly interested in Nepal's freshwater policy. The political relationship between Nepal and Bangladesh has strategic importance for India, given that both Nepal and Bangladesh have a strong anti-Indian lobby and from time to time have adopted anti-Indian stances on regional issues.[74]

The effects of climate change on Nepal's economy and the environment are enormous. Expected temperature changes, and retreating and melting glaciers could wreak havoc on Nepal's rainfall patterns. These will affect rice productivity, other agriculture, and the lucrative tourist industry. Lower agricultural productivity will deepen the food crisis, mostly affecting the poor.

Further, as the snow gathered by glaciers lessens, the freshwater flow to its Nepal's rivers, tributaries, and lakes will also decrease. Nepal has a large number of glacial lakes and these are becoming a serious environmental hazard. The Tsho Rolpa glacial lake near Kathmandu, which had an area of 0.23 km^2 in 1950, now has an area of 1.7 km^2. There are some twenty-six such potentially dangerous glacial lakes in Nepal. The United Nations warned Nepal in 1998 that twenty big glacial lake expansions could trigger huge loss of life and property, both in Nepal and in the downstream countries, particularly India, which is a lower riparian country to Nepal.

India signed three water-related treaties with Nepal: the Koshi Agreement in 1954, the Gandak Treaty in 1959, and the Mahakali Treaty in 1996. The first of these is due to expire in 2053, but the second has no specific sunset clause, and the third will remain in force until 2071.

Most Nepalese consider that the Koshi Agreement created difficulties rather than opportunities for Nepal. Nepalese activists complain that when India built a barrage on its soil to stop the Koshi River flooding, India shunted the problem upstream and it has now become Nepal's sorrow. The Koshi and the Gandak agreements, instead of contributing to the development of good relations between Nepal and India, have caused serious strains in that relationship. Subsequent amendments to the two treaties have addressed some Nepalese concerns. If either party feels that it is not serving their interests, it can ask for termination of the agreement.

There are provisions in the Mahakali Treaty for review every decade, or earlier if either country desires. The treaty requires having a binational Mahakali River Commission, guided by the principles of equity, mutual benefit, and no harm to either party. It focuses on the integrated development of the Mahakali River that includes the Sarada Barrage, the Tanakpur Barrage, and the proposed Pancheswar project. The treaty is in operation now.

Individuals in both countries have attacked some provisions of these treaties. For example, Medha Patkar, an Indian social activist and former commissioner of the World Commission on Dams (WCD) reprimanded both the Nepalese and Indian governments for signing these treaties without addressing the concerns of their respective citizens.[75]

In 1990 the Nepal Parliament passed a law that requires Parliament's approval by two-thirds majority for all treaties involving the use of Nepalese natural resources by foreign entities or which adversely affect Nepal on a long-term basis. Most accept that this law was directed at India, as its armed support to Bangladesh on self-determination grounds made Nepalese apprehensive about India's intentions. The law fueled mistrust and curtailed good relations between the two countries. It became fashionable for all Nepalese commentators to blame India for playing a big brother's role.[76]

Nepal has hydropower potential of around 80 gigawatts: one group favors large dams, while the other favors minihydro projects.[77] The minidam proponents won the policy debate against the global dam industry and the World Bank–supported dam project. The production cost difference per kilowatt between the small and mega projects is vast: US$700 for the small hydro scheme compared with an estimated US$5,000 for the large dam.[77a]

The Indian focus on jointly building large dams with Nepal may become a contentious issue with the new political environment in Nepal.[78] The possibility of constructing a number of small dams instead of large ones remains, should future governments decide to follow that path. The impact of any dam building on India's freshwater and energy policy would be substantial.

China has invested heavily in the infrastructure sector, including building roads across the country and up to the Indian border.[79] China has also settled its border disputes with Nepal by forgoing claims on ten villages after initially claiming eleven villages. Such actions inspire China to take a position of her own choosing when dealing with India on issues of critical importance to China. India disliked many Chinese investment decisions, as it thought these undermined India's special relations with Nepal.[80]

Nepal's political environment is changing rapidly. Nepal promulgated a new constitution in 1990, but could not develop the institutions essential for a functioning democracy. The shape and direction of the current Interim Constitution of Nepal, which came into effect on January 15, 2007, in the postmonarchy era is as yet unknown. China is cultivating its relations with Nepal assiduously. From a strategic point of view and policy perspective, India must consider all options within the overall regional and global policy precinct to ensure the country's relevance in Nepal's future water management development policies, and the impact on India.

Postrevolution developments have created tensions between caste and ethnic groups. Nepalese academic Krishna Khanal has argued that because there are so many different ethnic and case groups in Nepal, it is only possible to find three that have a clear identity in a federal Nepal.[81] In the savannah and grassland of Terai alone there are more than twenty factions.[82] The Maoist movement remains a loose federation of political activists committed to broad policy objectives, but it differs on the practicalities of being in government. The political disunity within Nepal's leftist movement has caused them to fail in the past.

The Terai people, known as Madhesi, maintain traditions and social and cultural customs that are similar to those of the Brahmins in India. India is concerned that a conflict in the region could create tensions that may entangle India in Nepal's internal strife. In 2008 the Madhesi blocked road access to Kathmandu, resulting in severe shortages of daily essentials that travel the route into Nepal from India.[83] India had warned Nepal that the Constituent Assembly election would not be possible unless the Madhesi issue were resolved through dialogue involving all parties,[84] but Nepal went ahead with the election and the Maoists became the single largest party in the Constituent Assembly.

Following the election Prachanda, the leader of the Maoists, became prime minister of a coalition government. He was forced to resign (after declining to step down voluntarily when President Ram Baran Yadav refused to back him against the army chief) after he declined to implement the government's position on the recruitment of army personnel.[85] The Parliament appointed a new government, but the Maoist leaders, Prachanda and his deputy Baburam Bhattari, did not join the new government.

Prachanda questioned the morality of the political parties and said that his decision to sack the army chief was to maintain civilian supremacy of the government. Earlier he accused India of interfering in Nepal's internal affairs during the standoff with the army chief. More concerning is his reference to the neighboring countries. Prachanda stated he resigned because he tried to see the country's relations with neighboring countries from a new angle, and thought that the Maoists should maintain good relations with India and China. Prachanda has maintained his rage against India, accusing it of openly interfering in Nepal's internal affairs.[86]

Sections within the Indian government consider that the Maoists in Nepal may provide a psychological boost to Indian insurgent groups, which are increasingly gaining ground in India.[87] The state government of Maharashtra publicly expressed its fear of the Nepalese Maoist involvement in the insurgency movement in the state.[88] The Indian government is concerned that the success of Maoists in Nepal will bring political instability, particularly in the northeast.[89]

Nepal is constrained by its geography, size and location. Nepal needs to protect its interests by adopting a cooperative policy with its giant neighbors in managing its water resources and by ensuring that all dealings are transparent and benefit its citizens and the region. Should there be a change in the political makeup in Nepal, it may ask India to renegotiate some or all of the existing water management treaties.[90] This remains the greatest strategic concern and security issue for India within the parameters set by the water crisis and climate change.

The Constituent Assembly is currently drafting a new constitution. The country has again fallen into political turmoil, as the Maoists called for an indefinite general strike all over the country, and demanded the current prime minister give way to a national unity government headed by the Maoist leader. The government has refused to accept this demand. The situation, both in political and development terms, does not augur well for either Nepal or India.

China's relationship with Nepal is friendly and it is also contributing significant resources to Nepal's development. The Ganges River receives waters from tributaries originating in Nepal. China's good relationship with Nepal may tempt it to pursue a policy to benefit China at the expense of India by playing the China friendship card or the Indian hegemony card.

Historically and culturally, Nepal has strong links with India, but events over the last few decades have cast shadows on the sustainability of this relationship. Insistence by Nepal and Pakistan at the 2007 SAARC summit that Afghanistan's admittance to SAARC is conditional on giving observer status to China is an indication that it would not be adverse in using the China card in its dealings with India. The strong political position of the Maoists in Nepal's political mosaic enhances such a possibility.[91]

ISLAMIC REPUBLIC OF PAKISTAN

Pakistan came into existence in 1947 as a separate homeland for the Muslim population. A large part of Pakistan is arid and deficient in rainfall. Rivers or waterways do not crisscross the country. The annual rainfall is over 1,200 mm in the Himalayan submountain region, and less than 100 mm in the south. Evaporation rates are extremely high, ranging from 1,300 mm in the northern Punjab to 2,800 mm in Sind.

Pakistan's main river is the Indus River, and four of its tributaries flow through Pakistan before entering the sea. It owns about 56 percent of the basin and shares its water with India under a World Bank–mediated agreement in 1960. The water flow from the Indus River often becomes a trickle when it enters the Arabian Sea.

Pakistan has seven major lakes, spread over four states. The aquifer extends to 300 meters over most of Pakistan. Overexploitation of groundwater without adequate replenishment has created problems.[92] Water tables have been falling in Punjab, Balochistan, and the Frontier Province. People from Balochistan have already started moving out of many villages. There are indications that by 2010, Quetta, the capital of Balochistan, will have exhausted its water supply.[93]

Irrigation uses 96 percent of Pakistan's freshwater, but incurs huge delivery losses, estimated at 40 percent on average. The Indus River basin is complex, and utmost care is required in the planning and execution of irrigation plans for this area. Researchers found that such care has been lacking, although the basin plays a very important role in Pakistan's economic prosperity. They commented that planning and designing irrigation systems is one thing, but their implementation and the water distribution is quite another.[94]

Annually, Pakistan harvests about 72 percent of its renewable water resource of about 234 km^3. Current projections indicate that yearly per capita internal renewable water resources will decline to 721 m^3 in 2050 compared to 1,756 m^3 in 2000.[95]

Water scarcity has reached a critical stage and the problems vary widely between the regions. Salinity and waterlogging have become a major problem in intensely irrigated areas.[96] The Punjab and some other parts of the country are suffering from gradual deterioration in soil capacity and increased salinity.[97] In 1993, senior official Syed Ayub Qutub expressed concern about the survival of Pakistan's land and the people because of the country's water deficiencies when saying, "I know of nine-year old farm [hand] lands in jail for murder over water disputes."[98]

Pakistan is facing a severe internal displacement problem and an influx of Afghan refugees. This is one of the largest humanitarian disasters Pakistan has ever faced. Large parts of Pakistan have been destabilized by the presence of the Taliban and other jihadi groups. Because of this, the size and magnitude of the future displacement levels remain uncertain at this stage.[99] Even the BBC found that ordinary people in the region speak of the future with trepidation, as the state was losing its grip and control of the people.[100] This has security and policy consequences for both Pakistan and India.

Pakistan's internal instability is of its own creation and is compounded by a curious mixture of events that followed its establishment in 1947. The death of Muhammad Ali Jinnah on September 11, 1948, just over a year after Pakistan's creation, the Cold War–led political intrigues, and the failure of the intelligentsia, and civil and military leadership in one way or the other contributed to this predicament. Religious affinity as a basis for state unity failed when Bangladesh seceded in 1971.

Democratic institutions never gained a firm footing in Pakistan. Consequently, peoples' anger and frustration did not find a legitimate vent working as a buffer between the citizenry and the government, because on each occasion when elected governments came to power, the army or their backers acted against them. A new constitution after each coup became the norm, and in the process the military became one of the natural contenders for power.[101] Electing a government is one thing, but to stick to democratic principles requires widespread commitment including political parties and their leadership. The turmoil in the streets of Pakistan stems from a mercurial mix of history, religion and politics, with explosive results.[102] Benjamin Disraeli argued in 1866 that "individuals may form communities, but it is institutions alone that can create a nation."

Notwithstanding this, Pakistan used the religious card to encourage the Mujahideen to fight the Soviets in Afghanistan, directly contributing to the growth the Taliban, an entity that is now even threatening Pakistan's very existence. Following the defeat of the Soviets, former fighters are now asked to fight their former comrades in arms, although Islam remains the binding force between them.[103] This contradiction has created a new militancy in which confused and frustrated young tribe members have been driven to embrace Islam as a cure for all their problems.[104] The apparent strength of the Taliban has contributed to the spread of the concept of a "cosmic war" that is practically impossible to contain.

The military dictators cleverly used this environment to benefit them and their cohorts. Following the September 11, 2001, attacks on the United States, the folly of pursuing such poor and outdated governance policies became apparent to many earlier backers of Pakistan. As Robert B. Oakley, a United States official, has noted: "Pakistan is beset by a whole bunch of chickens that are coming home to roost." Further, Steven Weisman has argued that the fundamental problem has been that "Pakistan's army now presides over a country forced to the edge of war against itself. Everything is contested: the end of ideology, the limits of Islamism, the rules of constitutionalism, the contents of the federation and the meaning of citizenship have all been compromised."[105]

Former Pakistan president Pervez Musharraf's use of emergency powers in sacking the judiciary to ensure his political survival was another contributing factor in the situation getting out of control.[106] With this development, Western pressure was alleged to have forced him to accommodate the democratic forces in the government. Benazir Bhutto, twice prime minister (1988–1990 and 1993–1996), was assassinated on December 27, during her reelection campaign, the mystery of which still remains unsolved. In the meantime, fundamentalists are continuing on whatever they do best. Many doubted the sincerity of Musharraf and some even claimed that the very existence of Musharraf contributed to galvanizing the movement.[107]

Many, including Hina Jilani, a prominent Pakistani attorney, who is also United Nations' Special Envoy for Human Rights thought that the 2008 election would be heavily rigged.[108] Consensus has been that the election was broadly fair, but Fatima (Bhutto's niece) accused that, of all parties, her aunt's party acted most fraudulently. Given that Musharraf used every trick up his sleeve to stay in power, it is difficult to accept that his administration would have allowed his opposition to twist the election against him, as he did not hesitate to sack the entire highest court of the country to establish the legitimacy of his own position.

At the same time it is also true that the history of governance in Pakistan has never followed a linear path. It has been mostly dictated by the whims of the army, in alliance with the bureaucracy, landed gentry (some eighty landed gentries control about one-tenth of the cultivated land in three of the richest provinces in Pakistan), captains of industry, and the legal fraternity.[109] The intrigue of international politics is such that those who have vowed to fight terrorism and raised their voices in support of democratic institutions only raised their collective muted opposition each time the army took control of the Pakistan government and on occasions they prefer to trust a military dictator to anybody else.[110] The *Economist* commented that they "are loath to do anything that might jeopardise their links with Pakistan and its intelligence services."[111]

The Pakistan Army is aware of this undercurrent in international circles. An *Asia Times Online* report that a former army dictator and president was suggesting in the press meet that he could stage a comeback of sorts. This assumed significance, as former prime minister (and now president) Asif Ali Zardari's popularity within his own party was waning. Many question his authority and ability, as Prime Minister Yousuf Raza Gilani draws closer to the army.[112]

Collectively, these factors have led back to the fundamental issue of governance, and Pakistan's structure and democratic values. In Pakistan, the system resembles a hereditary feudal dynastic democracy. Benazir Bhutto nominated her son, a student, to lead the party. Bilwal Bhutto Zardari (her son) may become a brilliant leader, but hereditary nomination should have no place in a democracy and other pluralist forms of government. By comparison, in Bangladesh and India, major political parties operate along feudal lines. The only small difference in India is that, despite the pillars of democracy not being strong, it has a history of some semblance of democratically elected governments. In Bangladesh the situation has slightly improved in recent years. For Pakistan, neither the Koran nor Sharia law offers much justification for democracy or civil society parameters. Instead, there is a rationalization of top-down and authoritarian decision making.[113]

Pakistan has also suffered from separatist movements in the resource-rich Balochistan, and internal strife such as conflict between *Mohajirs* (Muslim immigrants from India) and locals in Karachi, and water-sharing disputes between the Punjab and the Sindh. The Mohajir issue is complex, a clash between locals and immigrants in economic and cultural arenas. In 1995 alone an estimated twenty-five hundred people died in Karachi during the riot between these two communities. Independent estimates indicate that some thirty thousand to forty thousand Mohajirs were languishing in Pakistan's jails in the 1990s.[114]

Commentators have long argued that unless Pakistan reshapes itself from top to bottom, there is little likelihood of any fundamental change.[115] Pakistan's internal problems of ethnic divisions, religious violence, economic conflicts, and the situation created by the fundamentalists have worsened since 1983. Once religion becomes an integral part of the political equation, most policy decisions tend to lose their integrity, consistency and fundamental validity, as the history of Northern Ireland demonstrates.[116] One political commentator wrote, "Where once it [religion] used to be able by its total command of a worldview, to prevent the emergence of rivals, it can now only impede and retard—or try to turn back—the measurable advances that we have made."[117]

NOTES

1. Ekman 2010.
2. "Climate Change Needs to Be Discussed in 'Meaningful' Way: India," *Times of India*, April 28, 2010.
3. http://www.saarc-sec.org/userfiles/StatementofthePresidentofMaldives.pdf.

4. See S. M. Burke, "The Post-war Diplomacy of the Indo-Pakistan War of 1971," *Asian Survey* 13, no. 2 (1973): 1036–49.

5. "US, India Military Brasses Discuss China's Rising Power," *Times of India*, April 14, 2010.

6. LaPorte 1972.

7. Pant 2010.

8. Sperling 1998: 5.

9. Sabur 1995: 16.

10. World Bank 2006: 13, 163–64, 175.

11. Ray 2008a.

12. *Financial Times*, April 20, 2007.

13. Economy and Liberthal 2007.

14. Kahn 2007.

15. Haugen and Boutros 2010: 53.

16. Calvocoressi 2009: 495.

17. Landay 2008.

18. Gall 2007.

19. Baker 2007; *BBC World News*, May 19, 2006.

20. Gupta 2010.

21. World Bank 2005.

22. South Asian Association for Regional Cooperation secretariat, http://en.wikipedia.org/wiki/South_Asian_Association_for_Regional_Cooperation (accessed February 15, 2011).

23. Statement to that effect by the government planning secretary on February 26, 2004, during the release of the "Poverty Map" by the Bangladesh Bureau of Statistics and the UN World Food Program, Dhaka.

24. Chamie 2004.

25. Figures are purchasing power parity adjusted.

26. Income distribution figures relate to 2001 and poverty level figures refer to 1995/1996 fiscal year estimates.

27. Badruddoza Chowdhury, a former president, in an interview with the AFP in February 2004: "Ex-president Warns Bangladesh Could Become 'Failed State,'" http://story.news.yahoo.com/news?tmpl=story&cid=1534ncid=1534&e=7&u=afp (accessed February 23, 2004).

28. "Politics, Corruption Damage Bangladesh's Image"—British envoy in a statement on March 24, 2004; http://story.news.yahoo.com/news?tmpl

29. United Nations 2007: table 21.

30. Grynberg 1998: 6, 11, 38, 41; Gounder and Xayavong 2001: 5; Read 2001: 30–31; Arthur 2001: 603; and de Jonquieres 2004.

31. "The Tenacity of Hope," *Economist*, January 3, 2009.

32. Mustafa 2007; Fair 2007.

33. "Interim Government Launches Onslaught on Bangladesh's Dynastic Political System, Exiles Two Top Political Leaders," *Global Insight*, April 20, 2007.

34. An interview by ABC journalist Bill Bainbridge with Sheikh Hasina and I. S. Chowdhury, editor, *Bangladesh Observer* in the Asia Pacific program on April 23,

2007; http://www.radioaustralia.net.au/asiapac/programs/s1904708.htm (accessed April 24, 2007); Rahman 2006.

35. Ahamed 1998: 116.

36. Rahman 2007.

37. "Farakka Barrage," *Statesman*, October 30, 2004; Crow 1995: 26–54.

38. Ramakrishnan 2004.

39. *The Hindu*, October 11, 2004; Shiva 2003: 66.

40. Reuters-Yahoo news, October 31, 2004; *Daily Star* (Bangladesh), 4/260 and 4/283.

41. "Stemming the Flood," *Far Eastern Economic Review*, October 23, 1988.

42. Sobhan 2000.

43. Chapman 1995b: 184.

44. *Economist*, October 9, 2004.

45. Gopinath, Dasgupta, and Surendran 2003: 13.

46. *Statesman*, March 27, 2009. The Indian print media reported that the security forces in Bangladesh are investigating alleged links between a UK-based charity and an Islamic school after a raid uncovered what official described as a terrorist training facility.

47. BBC, "We Are Very Afraid after This," *World News*, March 28, 2009.

48. Chari 2003: 33; Hussain 2003; Hussein 2004.

49. Vijay Thakur, "Influx of Migrants Alarms Intelligence Agencies," *Statesman*, February 3, 2008:

> The Intelligence agencies have raised an alarm over the influx of illegal migrants from Bangladesh. . . . It seems that the Intelligence agencies of our neighbouring countries [are] adopting the same strategy vis-a-vis the Northeast, what it had done in Kashmir. Secondly, Northeast insurgency is becoming a cocktail of the activities by the region's insurgent groups and Islamist militant outfits from Bangladesh. . . . Inputs from Bangladesh, have suggested that both are getting arms and ammunition from the same source and also indirect help from certain forces from Bangladesh.

50. Government of India 2004: 10.

51. Chapman 1995a: 20.

52. Bhutto 1969: 163. Sheikh Mujibur Rahaman wrote, "Because Eastern Pakistan must have sufficient land for its expansion and because Assam has abundant forests and mineral resources, coal, petroleum, etc., Eastern Pakistan must include Assam to be financially and economically strong." Cited in "Report on Illegal Migration into Assam," prepared by the governor of Assam in his report to the president of India in 1998. For further details see Hussain 2003: 131, 148–49.

53. United Nations Conference on Trade and Development 1989: 39.

54. Fair and Ganguly 2007.

55. Fair 2007.

56. Bowring 2009.

57. Chapman and Tompson 1995: 187.

58. Myers 1996: 20.

59. Sivaraksa 1993: 256.

60. Given the recent record of China referring to historical incidents in relation to its national border, one wonders whether China, in future at a time of its choosing, may question Bhutan's sovereignty.

61. Dorgi and Pradhan 1987.

62. United Nations 2007/2008: table 2.

63. United Nations 2007/2008: table 8.

64. "PM [Prime Minister], Bhutan King ink updated friendship treaty"; external affairs ministry spokesperson, New Delhi, February 8, 2007.

65. Chapman 1995a: 20; Bakshi 2004: 274–75.

66. Majumder 2008.

67. Sherpa 1994: 9, 72.

68. Dahal 2001: 93.

69. Bray, Lunde, and Mansoob Murshed 2003: 107–32.

70. United Nations 2007/2008: table 8. National estimates are based on population-weighted subgroup estimates from household surveys. Population below $2 a day is the percentage of the population living on less than $2.15 a day at 1993 international prices (equivalent to $2 in 1985 prices, adjusted for purchasing power parity). Poverty rates are comparable across countries.

71. Upadhyaya and Sapkota 1987: 98.

72. Besides three major rivers—namely, Karnali, Gandaki, and Sapta-Koshi—other rivers are the Mechi, Bakra, Kankai, Lohendra, Budhi, Rato, Bagmati, Kamal, Tinau, East Rapti, West Rapti, Babai, Khutiya, and Mohana.

73. United Nations 1986.

74. "Stand By for the Great National Unraveling: Foreign Hand," Nepali Times, February 9, 2007.

75. *Kathmandu Post*, December 11, 2002.

76. Iyer 2003: 223.

77. United Nations World Water Development Report 2003: 262.

77a. McCully 1996: 230 and 270.

78. Bell 1994; McCully 1996: 230.

79. Nath 2004: 65–66.

80. Dutta 1998: 101.

81. Khanal 2007.

82. *Economist*, November 10, 2007.

83. "Nepal: Protests Cause Fuel Shortage in Capital," http://www.abc.net.au/ra/asiapac/programs/s2168213.htm (accessed February 20, 2008).

84. "Resolve Madhesi Issue before Polls, India Tells Nepal," *Statesman Weekly*, February 9, 2008.

85. "Maoists Sack Army Chief, Nepal Teeters," *Statesman*, May 4, 2009, http://www.thestatesman.net/page.news.php?clid=1&theme=&usrsess=1&id=253115.

86. "Nepal Maoists and House Blockade," *Statesman*, May 23, 2009; "Nepal Crisis," *Statesman*, December 22, 2009.

87. *Statesman Weekly*, February 12, 2005. A defense ministry spokesperson, in a press conference on February 9, 2005, acknowledged that close links exist between Maoists in these two countries, India and Nepal.

88. "Nepal Hand in Naxal Strike?" *Statesman*, May 23, 2009.

89. A group of eight hundred Nepalese who were alleged to be employed by the government of India on a contract basis took this fluid situation to accuse the Government of India that it was not giving them their due pension. They were hoping that in such a situation the Government of India would feel embarrassed and perhaps buy them out by making some payments to them; http://www.thestatesman.net/page.news.php?clid= (accessed June 9, 2004).

90. Khadka 1994: 344–94.

91. R. Mishra 2007; Haviland 2007.

92. Ghassemi, Jakeman, and Nix, 1995: 371–78.

93. http://www.irc.nl/page.php/405 (accessed February 10, 2004); Clarke and King 2006: 23.

94. van Halsema and Vincent, 2006: 55–91.

95. Ghassemi, Jakeman, and Nix 1995: 94–99).

96. Kahlown, Khan, and Azam 2006: 35–54.

97. Jensen, Rangeley and Dieleman 1986: 32.

98. Simon 1998: 69.

99. Jill McGivering, April 23, 2009. A BBC journalist, after meeting the displaced families in the bleak camps in the North-West and wealthy Pakistanis in Islamabad about their experience and fears for the country's future reported on "how Islamic extremism is affecting Pakistan society." She also witnessed firsthand the audacious and bloody siege of the Lahore police compound by the Islamic militants. Report is available in the BBC website (see McGivering 2009).

100. http://news.bbc.co.uk/go/pr/fr/-/1/hi/programmes/from_our_own_correspondent/7980735.stm (accessed April 4, 2009).

101. Rizvi 1998.

102. Robinson 2007.

103. Norton 2004: 131–32.

104. *New York Times*, November 8, 2004.

105. Newberg 2008.

106. Santhanam 2007; Anthony 2007.

107. Akhtar 2007: 73–79.

108. Parameswaran 2007.

109. Alavi 1983.

110. See Haqqani 2007, 2008; Phillips 2008.

111. *Economist*, November 10, 2007.

112. "Pak Army Warns Zardari," *Statesman*, March 10, 2009.

113. Wiarda 2003: 111.

114. Singh 1998.

115. Ali 1983:195.

116. Also see Rhode 2007; Rashid 2007a, 2007b, 2008; Khan 2008; Yas-Meen 2007; Gordon 2007; Chanda 2007a; Newberg 2007.

117. Hitchens 2007: 282–83.

Chapter Seven

Copenhagen: Where to Now?

The Copenhagen Summit has ended, but most countries, environmental activists and informed commentators have been vocal in their claims that it did not produce the desired outcome. There was much excitement and expectation that the post–Kyoto agreement would be signed in 2010 and would set country targets for reducing CO_2 emissions in the atmosphere. This has not happened despite the collective wishes of the global community and knowing that the consequences of the world getting hotter would be disastrous for all, but particularly for poorer nations. One commentator, Tim Alfred, summed up the situation perfectly: "Rarely have verdicts on international summits been quite so united in their expressions of disappointment, derision and outright condemnation."[1]

Others, although profoundly dissatisfied with the outcome, were less critical. Harvard University's Rob Stavins made the point that even participating countries could not agree to the Kyoto Protocol; many signatories did not achieve the Kyoto targets, and did not provide any indication that they would endeavor to do so.[2] In that sense, the Kyoto Protocol was a dead agreement. In short, Stavins' view was that given the rancor displayed at the Copenhagen Summit, the climate change policy process should best be viewed as a marathon and not a sprint. Certainly there were some basic flaws in the global expectation that the summit would achieve lasting results. Still, depending upon details yet to be worked out, the summit could well turn out to be a sound foundation for a portfolio of domestic commitments, which could become an effective bridge for the longer-term arrangements between countries.

A number of climate scientists have gone further. They consider that even the emission target set by the Copenhagen Accord in December 2009 will fall short of achieving its objective of limiting any temperature rise to 2°C. Many scientists now argue that, to meet the temperature target, the emission target has to be 48 gigatons of CO_2, and not the agreed target of 40–44 gigatons.[3] Another group of scientists argues that although climate change is the greatest collective problem in human history, the approach followed to achieve an agreement was made harder than necessary. They argue, "There is no alternative to negotiating treaties to address climate change, but there is an alternative negotiating strategy. A better way to negotiate [is] to break this colossal problem up into smaller pieces, addressing each piece by addressing the best means appropriate."[4]

The scientists argue that the differences between countries are vast and one size does not fit all. For example, rapidly developing countries (such as India, China, South Africa, and Brazil) refuse to agree to an emission target that puts them on an equal footing with industrialized countries, while highly industrialized countries such as the United States refuse to abide by any emission agreement unless these countries commit to reduce their CO_2 emissions.

When it comes to strong national economic and political interests, countries do not hesitate to flout those norms and standards that they expect other countries to follow. Brazil, a regional leader and an important player at the Copenhagen Summit, recently did not hesitate to hastily proceed with the construction of Belo Monte, the world's third largest dam project—a hydroelectric complex on the Xingu River in the state of Pará—after beating back a last-minute court injunction and suspension order. The opposition to this project came from all walks of life, but particularly from environmentalists and celebrity personalities such as film director James Cameron (the director of the film *Avatar*), who were drawn into the "native versus exploiter" theme. The project is estimated to displace over forty thousand families, of which a significant number belong to Brazils' Indian tribes.[5]

The problem appears to be that, with the growing economy, the demand for power is growing exponentially. Interest groups and lobby groups are known for their capacities to manipulate inputs that are favorable to their interests, and oftentimes an outcome will indicate that the project failed to keep a balance between private and public interests. Two books published in 1990s are outstanding examples of such practices.[6] The lamentable part in all of this is that professional bodies, government officials and politicians, and even multilateral agencies often knowingly or unknowingly become parties to such arrangements.

The World Commission on Dams, an independent international body established in May 1988, whose report "Dams and Developments: A New Framework for Decision-Making" was published in mid-November 2001, brought together an array of experts from all sides of the argument who understood the cost of dams to the local population and environment, as well as the national benefits; it was ignored in Africa. Projects are rushed; huge contracts are open to corruption, and engineering can be substandard. Graft and incompetence have meant that many dams (whether in Africa or elsewhere), once built, have failed to live up to the promise. Even some donor agencies are not immune from such practices.[7]

The country delegates were far more disapproving of the failure of the Copenhagen Summit to achieve any realistic goals than individual scientists and commentators. South African president Jacob Zuma said that the summit declaration was unacceptable, while another member of the "African Union" delegation added that it was limited not only in terms of what it did to save the planet, but that the number of nations accepting the declaration did not extend beyond the twenty-eight represented at the late-night negotiations.[8]

Collectively, African countries argued that a global agreement must ensure that the temperature in Africa does not rise by more than 2°C. The Ghanaian envoy observed that if one talks of a 2°C rise globally, then certain places in Africa will exceed that mean. A recent study suggests that the real long-term increase will actually be between 3.5°C and 4.5°C.[9] More concerning is the comment from another powerful African delegate, Kamel Djemouai from Algeria, who added, "I can assure you that no one will have any chance to divide African countries. We are going to stay united and unified as African countries, and we are going to move as one Africa in this process."[10]

If the Africa caucus remains steadfast in its position and insists on an acceptance of its baseline temperature in the final resolution, then the chance of attaining a universally legally binding decision at the sixteenth conference of the parties to the Kyoto Protocol, to be held in Cancun, Mexico, between November 29 and December 10, 2010, appears remote.

The spokesperson for the G77, the Alliance of Small Island States, and a number of Central American and South American caucus members also voiced their indignation at the outcome. A spokesperson argued that it would throw Africa "into the furnace" and compared the rich countries' heartlessness to the genocide by Adolf Hitler.[11] One Peruvian participant along with others from the region felt that such summits were more of a power game rather than about meeting real needs of countries.[12]

US president Barack Obama agreed that people were justified in being disappointed about the outcome, but defended that, at least, the summit attendees were able to agree on nonlegally binding targets for all countries.[13] Indian prime minister Manmohan Singh, a member of the drafting group, also indicated his dissatisfaction.[14] Disagreement within the Indian core policy group was intense. Consequently, three senior members of the group have either resigned or not been recommended for renomination in the group.[15]

Indian commentators varied widely in their assessment, from outright condemnation to mild support. One veteran journalist from India, Rajinder Puri, dubbed that country's participation at the high table a betrayal and commented that India was becoming Beijing's puppet and that the Indian government had taken care of any security threat from China.[16] One former foreign secretary took an opposite view and argued that, although the summit outcome had been vague, from India's point of view the display at the summit was an important new manifestation of its growing international weight where it played its full part in framing the decision.[17] Public reaction was vocal and some people used mock funeral processions to condemn the outcome.

Copenhagen Climate Change Council chairman Tim Flannery watched the summit proceedings closely and observed, "One of the most interesting things that happened . . . has been a real rift between the poorest countries and the island nations and those semideveloped countries such as China and India. . . . I'm astonished that . . . the G77 has held together for so long."[18]

Three main issues in the broad nonbinding agreement were that

- global warming should be limited to less than 2°C (or 3.6°F)—the degree of warming generally accepted as being "dangerous";
- rich countries must register the emission cuts they will make by 2020 by the end of January 2010, although there is no guarantee that this will limit future warming in the future (as what countries announce they will cut is left up to them), and many (including UN officials) consider that the date should be considered a flexible date; and
- between 2010 and 2012, rich countries will provide US$30 billion to poorer countries, with the goal of providing an annual sum of US$100 billion by 2020.[19]

CHANGING GLOBAL ALLIANCES AND SHIFTING POWER BASES

The global political and economic alliances have been shifting in the post Cold War period. Rapidly rising economies have forced the pace of such change. New relationships are emerging which were unthinkable a couple of decades ago. Everywhere this is true, but more so in Latin and Central Americas. The Caribbean nation of Cuba was once the only significant thorn in the side of the United States. Now countries such as Bolivia, Nicaragua and Venezuela are adopting a strong, independent line in their economic and foreign policies, and forming new defense and economic alliances. Brazil, the largest of the South American countries, is asserting the continent's economic and political independence. For example, between February 28 and March 5, 2010, US secretary of state Hillary Clinton visited a number of Latin American countries to garner general support for the US position on important international issues such as stronger UN sanctions on Iran. Clinton was extended a courteous hearing, but she fell short of extracting a firm commitment from most of the states that mattered.

Venezuela signed an arms and energy deal with Russia, which filled the gap left by the US arms embargo. The deal complicates US efforts to improve relations with Venezuela,[20] and undermines the former's traditional authority in South American affairs. Such an agreement would have been unthinkable in earlier decades.

Globalization, the emergence of new markets, and the need to ensure a reliable long-term source of industrial and energy raw materials are prime movers in forming new alliances. The days of permanent relationships are long gone; increasingly such relationships are dictated by economic, security, and political realities. These alliances are yet to be tested in terms of durability, but the increasing assertiveness of the hitherto weaker nations on many international issues cannot be ignored in the rapidly changing international political and economic environment.

One example is the new alliance in the American continent in 2010 known as the Community of Latin American and Caribbean States. Once its charter is developed, the grouping will be formally established in July 2011. It includes Cuba, but excludes the United States and Canada. Many consider the organization to be a potential rival to the Organization of American States (OAS), an institution established (in its current guise) in 1948 to combat communism and promote democracy and human rights.

The inaugural chairman of the new alliance is Mexico president Felipe Calderon, although this nation has long been a trusted US ally on international issues. At the inaugural meeting he is reported to have said, "The challenge that faces our region is not a matter of left or right; it is not a question of ideologies or doctrines. Rather, it is the alternative between the past and the future, a future where the values we believe in—democracy, justice and freedom—can flourish."[21]

The chairman is further reported to have said, "[We] must, as a priority, push for a regional organization . . . and promote the regional agenda in global meetings."[22] It is reported that the group will come into formal existence in 2011 and that the group's current meeting, although not open to the public, confirmed that it had discussed the issue of Argentina's dispute with Britain. If this is true, then it is clear that the new group will not hesitate to take a stand on politically sensitive issues that relate to their members' interests, even though the issue might be against the interest of their erstwhile European ally.[23] It is too early to predict the future of the new organization vis-à-vis the OAS. In view of the critical importance of the Amazon basin and the Brazilian rain forest in steadying CO_2 emissions and capture, and the rising sea level threat to many Caribbean island countries, this new alliance may play a key role in the Cancun summit in 2010, which may not necessarily please one or more of the powerful, industrialized countries. Rivalry aside, no longer can one assume the perpetuity of political alliances.

Since the days of the Bretton Woods system (which lasted from 1944 until the early 1970s and created the International Monetary Fund and the World Bank), global economic and political decision making has been monopolized by the annual meeting of finance ministers (or G7) since the first meeting in the mid-1970s. Only in the last decade has there been a shift, as the G7 nations (Canada, France, Germany, Italy, Japan, the United Kingdom and the United States) come to realize changes on the ground. A few have even argued that the G7's monopoly has been an anomaly. This was partly rectified with the first G20 Summit in 2008 and then with its confirmation in 2009 that, henceforth, it would serve as the body of global economic governance.[24]

Jean-Pierre Lehmann, a professor of political economy, has argued that the Copenhagen Summit was generally seen as a failure, but that it was humiliation for the European Union. The European Union did not play any part in the eleventh-hour drafting of the Copenhagen Accord, where the key players (besides the United States) were China, India, Brazil, and South Africa. The European Union's failure was that it had no effective caucus at the summit and, thus, no strategy. The European Union needs to understand the importance of shifting alliances in an interconnected and globalized world.[25]

There is some irony in this. In the entire climate change debate, EU countries adopted a positive role compared to other industrialized countries. Some may argue though that the European Union was not bold enough and point to the roles played by the United Kingdom, France, Germany, and the Netherlands. However, following the collapse of any legally binding agreement and the European economic crisis, the role of these nations is now subdued. For example, France has decided not to introduce a carbon tax given its high unemployment and the financial crisis, but remains committed to proceeding with the tax, if it is agreed at EU level.[26] Other countries such as Australia were forerunners to introduce a carbon tax, but even this continent has now abandoned the plan, citing the summit's failure to adopt a legally binding resolution and the failure of the Australian Parliament to pass the legislation.

The Global Economic Forum has argued in similar terms. In 2008, when it sent its climate policy recommendation to G8 leaders, the forum specifically noted, "We are committed to addressing climate change. . . . We also applaud and support the Bali Action Plan . . . to succeed the Kyoto Protocol. . . . The new framework must be environmentally effective and economically efficient. All the major economies must be party to it, including the United States, China and India.[27]

The Global Economic Forum's 2009 conference ended with slight apprehension. A panel from the forum argued that the gap in growth rates between of the developing and developed worlds was widening. Richer countries would invest more aggressively in emerging markets in order to maintain their own growth, which would be beneficial to the emerging world.[28] This has major implications for political leaders, who must ensure that they agree on an unambiguous international goal for overall reductions in greenhouse gas emissions.

The shift in the international power configuration appears inevitable, and the implications are complex. If history is any guide, this shift will certainly raise tension between nations as, generally speaking, no individual, community or nation wants to cede its power base or the capacity to influence the outcome of its chosen policies voluntarily, without a fight. This stance has become more evident over the last few decades.

Even former US president George W. Bush publicly acknowledged this in 2009. On the issue of India having a permanent seat at the UN Security Council (UNSC), he agreed that present veto-wielding countries were less than happy with the idea once they realized an expansion of the permanent membership would mean a dilution of their influence. He said, "The Security Council should be changed given the new realities of the world. . . . But the politics is difficult."[29]

All outcome assessments did not truly reflect the situation at grassroots level. Most countries and organizations expressed disappointment at the outcome of the Copenhagen Summit, but three assessments, in particular, defied the trend. UN Secretary-General Ban Ki-moon called the Copenhagen Accord a beginning. He ignored the achievements of the "UN Framework Convention on Climate Change" negotiated at the UN Earth Summit in Rio de Janeiro in 1992; the "Kyoto Protocol" signed in Japan in 1997,[30] and the Bali action plan that was agreed in 2007.[31] The Secretary-General thus ignored the achievements of the organization that he leads.

Second, Yvo de Boer, the then executive secretary of the United Nations Framework Convention on Climate Change (UNFCCC), thought that it was an impressive accord. This is directly in contrast to opinion expressed by both large and small nations. The three big issues on the table for consideration were legally binding emissions targets, verification, and funding.[32] Of these, a tentative agreement was only reached on the funding issue, with Japan offering US$15 billion immediately, the United States offering an undetermined sum (with a promise to raise US$100 billion a year by 2020), and China declaring that it would not seek any assistance from these funds.[33] The other two issues remain untouched, primarily because of the "never, never" approach adopted by China, in particular. Thus, the executive secretary's assertion that it was an impressive accord was misplaced.

The executive secretary, in his first formal press conference following the summit, however, did admit that it was not the agreement desired by the United Nations. He further said that the countries that attended the summit would have until the end of January 2010 to declare their position on the accord—an accord which, in any case, is not legally binding. The January 2010 deadline remained a flexible deadline.[34]

Further, Ed Miliband, former UK secretary of state for energy and climate change (current leader of the British Labour Party and head of Her Majesty's alternative government in the United Kingdom), singled out developing nations for not achieving legally binding targets.[35] He particularly singled out China.[36] This indicates that, first, he failed to understand the multiplicity of national objectives of nations at the summit; and, second, he should have realized that nations have their own policy priorities and that adopting a broad-brush approach is unhelpful. The Chinese Foreign Ministry spokesperson considered the minister's opinion as being designed to sow discord among developing nations and that the minister should have better known the mood of the summit participants.

A BROAD-BRUSH APPROACH DOES NOT MEET DIVERSE INTERNATIONAL NEEDS

Global warming is an all embracive issue, and it is difficult, if not impossible, to arrive at a consensus between countries as economically diverse as say Norway, with an annual purchasing power parity–based (PPP-based) annual per capita income of US$53,433 and the Democratic Republic of Congo, with a PPP-based per capita income of US$298.[37]

The draft paper prepared by the summit chair (the Danish foreign minister) upset the delegates from many of the developing countries. The approach adopted in the paper incensed them. In hindsight, it could be argued that this paper vitiated the mood of many participants, which possibly did have some bearing on the eventual summit outcome. The situation became so hopeless that the European Commission representative, Artur Runge Metzger, cleverly defused the situation by arguing, "There's no validity. It's just a piece of paper. The only texts that have validity here are those which people negotiate on. If she [the summit chair] wanted to introduce a text, or the Danish—the president—prime minister wants to introduce it, he will have to do it in a formal session. Certainly not through a leak in the press.[38]

The climate change issues are sensitive and require deft handling by all parties, particularly because of the sharp differences between industrialized and developing nations. But the summit chair's inept handling of the complex issue complicated matters further. The effects from climate change are universal, but the economic and technical capacities of nations to mitigate its impact vary substantially. A rational and acceptable outcome can be achieved through the cooperation of all countries.[39]

US president Barack Obama realized this. At the Copenhagen Summit, he scrapped Bush's climate legacy, although he had scant room to offer deeper concessions as he was gingerly steering an emissions bill through Congress. It is clear that the political and economic realities of nations shape their individual positions, and it is very difficult to dictate terms from the top, at government level.

The *New York Times* and the *Washington Post* also conceded that domestic political and social considerations from the economic downturn and associated uncertainties made it unlikely that any country was going to make real economic sacrifices for long-term problems, especially the two powerful countries in this regard—the United States and China.[40] The power of the emerging multilateralism, as observed in recent years, was unequivocally confirmed by the events at the Copenhagen Summit.[41]

The independent external verification issue was core to China's "never, never" stand. The veracity of China's objection needs to be understood in its historical context, with its technical, procedural, and political problems.

First, we need to assume that countries agree to independent external verification of whether or not the declared objective of a reduction in CO_2 gas emission has been achieved. So long as the world remains divided into various groups and subgroups, the question of which groups are acceptable as independent will remain debatable. The events leading to the Iraq conflict are examples. Even if this issue can be resolved amicably, the major problem is a technical one.

For example, when a country reports certain levels of emissions of various gases, the emissions must be measured at different stations to ascertain which way the wind is blowing at time of measurement, and take into account the weather conditions. Experience with such measurements shows that many countries are emitting more gases than first thought or reported. The problem is that it is difficult to replicate the conditions under which measurements are taken.

Thus, it not only becomes a costly exercise requiring the commitment of scarce resources by poorer countries, it also becomes a contentious issue even if the reporting country has scrupulously followed the guidelines. Scripps Institute scientists found that they had to fight the bureaucracy and budget cut proponents repeatedly to secure funding to monitor gas emissions. Monitoring, reporting and verification are no easy tasks, even for rich, industrialized countries. Further, it is completely understandable why some countries are reluctant to agree to allow outside monitoring; it will be easy for some measurements to produce a negative finding, although the country concerned may have reported its data in good faith.[42] Many nations are also concerned about their sovereignty, as it might be easy to undermine a country's reputation by referring to this mechanism.

The roles played by powerful nations through the years, while addressing global conflict and disputed issues; also provide grounds for emerging countries to oppose foreign intervention to measure their performance. For example, in a report on the final phase of the 1993 Uruguay Round General Agreement on Tariffs and Trade (GATT) held in Geneva, a journalist found that journalists were often privy to the deliberations of the world's superpowers, but not to those of ministers from the Third World, who were most affected by the decisions made in these meetings. Often they were forced to wait for hours, begging the emerging journalists to tell them the latest developments of any negotiations. Lack of appropriate representation by many poorer countries, inadequate research staff to prepare them for effective intervention in the complex deliberations, and lack of confidence to confront the larger players in the process are some of the reasons for developing a distrust of outside intervention.[43]

Further, even the relatively less powerful, industrialized countries publicly argue that using the political and economic power of powerful countries runs the risk, in the long run, of losing public support for any issues under deliberation. It is simply that many countries draw on their past bitter experiences.[44]

WHERE TO BEGIN AND THE COMPENSATION MECHANISM

The G77 countries wanted the Kyoto discussions to be on the table at the Copenhagen Summit. In particular, they wanted the Kyoto commitments to be a prerequisite for any future agreement. This was refused and the G77 countries walked out of the process.[45] This situation arose when attempts were made to close the chapter on the Kyoto Protocol and to make a fresh start. The ostensible reason for this initiative was to protect Canada from a large penalty payout. Canada has incurred a severe breach of its emissions targets under the Kyoto Protocol and owes about a US$1 billion penalty under the Kyoto Protocol. If Kyoto continues, Canada will owe US$1.3 billion.[46] Canada's emissions today exceed the level allowed under the Kyoto Protocol by more than 30 percent, and it has no plan to comply with the levels. Canada was not alone; some other developed countries also wanted the Kyoto commitments to cease and then to move on to the new commitment period with a clean slate.

The impasse between the industrialized and developing countries about who is to foot the bill for implementing the steps necessary to mitigate emission of harmful CO_2 is difficult to resolve, as it will largely depend upon the type and degree of economic and social dislocation suffered by individual countries, which need to be supported. Industrialized countries made a collective commitment of US$30 billion over the next three years to 2012, payable to the most vulnerable countries. The United States, the European Union and Japan have collectively agreed to pay a little more than $25 billion, the US pledge being contingent on Congressional approval. Further, they have committed to a goal of jointly mobilizing US$100 billion annually by 2020 from public and private sources.[47]

The agreement is not explicit on how much of this fund will be contributed by individual countries, nor does it specify any rules for spending the money.[48] Some are even apprehensive that the targeted amount is, at best, half of what is needed,[49] as confirmed by the experience of the Scripps Institution of Oceanography. It has been further argued that there was no guarantee that the agreement will come to fruition. The example of Kyoto Protocol was cited that some countries did not comply with their commitment, others complied only by some clever accounting, while yet others did not even bother.[50]

During past decades there have been a number of global summits addressing issues such as poverty alleviation, global food security, and setting millennium goals. In each of these summits many donor countries committed to contribute toward those causes. Many of those commitments remained unfulfilled. For example, in 2001 the European Union joined the United Nations (the Bonn Declaration in 2001) to set up two special funds to tackle climate change to which the industrialized countries promised to pay between US$1.6 billion and $2.87 billion by 2008. The Declaration, signed by 20 industrialized nations—the 15 countries that then made up the European Union, plus Canada, Iceland, New Zealand, Norway and Switzerland. In 2009 the BBC investigated whether the poor countries have seen all of the promised funds. It found that there was much confusion whether all the moneys were paid. It found only US$260 million has ever been paid into two UN funds earmarked for the purpose.

Dr. Marc Pallemaerts, who drafted the Bonn Declaration in 2001 when he was the deputy chief of staff for the European Union's Belgian presidency, admits some developing countries may have been led to believe the promised money would go solely into the UN funds. The Bonn Declaration was surrounded by confusion and has led to mistrust between developed and developing countries. Artur Runge-Metzger, the senior climate change negotiator, maintained that the EU had lived up to its end of the bargain, but could not provide data to show it did pay the money through bilateral and multilateral means. Boni Biagini (who runs the UN funds) believes much more should have been paid in.[51] A BBC investigative report on Liberia confirmed that Liberia urgently needed cash to deal with its climate change–induced crises that was putting towns and cities at risk of flood from rising sea levels. Richard Myungi, a climate change negotiator for the Least Developing Countries, lamented, "We feel frustrated, we feel betrayed."

If previous promises made by rich countries are anything to go by, the world's developing nations could again be shortchanged.[52] A. C. Revkin, a biologist and veteran *New York Times* environmental reporter, confirmed that, when one looks at this issue over history, there has been hardly any change. It is therefore unsurprising that poorer nations would be asking for a guarantee.[53]

There are two other issues that should concern the poorer economies, who require both financial assistance and transfer of technologies to mitigate the effects of climate change on their economies and societies.

First, assuming all industrial economies are genuinely interested in contributing to the fund as mooted at the Copenhagen Summit, the issue remains whether they will have any capacity to contribute in the fund. Australia aside, no developed economy has yet emerged from the 2007–2008 recession (Germany excluded, as it has achieved partial economic recovery). The US economy is showing some signs of recovery, but this recovery is tentative for now.

The European Union is in financial shambles; Greece is in an extremely parlous state, while Ireland, Spain, Portugal, (possibly) Italy and a few other nations are not far from it. International speculators are doing what they are best at, by raising the cost of funds that could make the position of smaller economies such as Greece (2 percent of the European Union's GDP) intolerable, and act as a tipping-point for metastasizing the other twenty-one EU economies which, in turn, will create a debilitating effect on the global economy. The fear remains that Greece's financial crisis could affect the economies of European countries which, in turn, may adversely affect the economies of the United States, the United Kingdom, and even the entire global economy.[54]

The European Union has been forced by circumstances to establish a trillion dollar fund in collaboration with the International Monetary Fund to prevent further economic fallout from the financial crisis. This crisis and other economic and political imperatives have drastically reduced the paying capacity of almost all industrialized economies. Funds are required to cushion developing countries to cover the negative consequences of any restriction on limiting their CO_2 gas emissions. The question arises as to who will assist to cover this resource gap in the event of countries agreeing to limit their gas emissions?

The financial capacity of industrialized economies to mitigate climate change is now drastically reduced, and in some cases nonexistent. The capacity of the United States to pay depends on the outcome of its draft legislation currently before Congress. Current indications are that the legislation is unlikely to pass by both the House of Representatives and the Senate. The reasons for this are simple. Robert Schiller, comparing the experience of US president Franklin Delano Roosevelt in the 1937–1938 recession following the Great Depression of the mid-1930s, argues that concerns about the national debt have started to emerge.[55] The current US deficit has already skyrocketed. In the post-Thatcher-Reagan era, deficit financing is frowned upon in most industrialized economies.[56] Schiller admitted that the Tea Party movement in the United States has assumed amazing proportions. The climate change–related public outlay became a scapegoat in this environment.

It is reported that the International Monetary Fund (IMF) is planning a $100 billion fund to help countries to mitigate the effects of climate change. It has asked countries to think "outside the box," as the ability of developed countries to contribute is limited due to their own financial crises.[57]

The climate change issue is critical, but it is also a long-term issue for most poor countries, as their daily struggle to survive becomes a priority issue. This is the very reason why financial support for mitigation is to be coordinated globally as far as is practicable. The IMF's managing director was unequivocal that traditional financing strategy will be insufficient to address the problem, and that the most important objective is to achieve political consensus.

ISSUES CONCERNING TECHNOLOGY TRANSFER

Another important issue in the climate debate is technology transfer. Despite numerous international commitments to promote the transfer of climate change–related technologies to developing countries, transfers are not occurring at a rate fast enough to assist them in mitigating and adapting to the effects of climate change. The technology transfer is required to limit gas emissions by reducing pollution, increasing the efficiency of production and energy usage or facilitating the capture and storage of carbon. The current intellectual property regime (IPR) is inadequate for an effective climate change–related technology transfer. There are many problems, but the critical one appears to be institutional in nature.

The recent trend in most industrialized societies has been to privatize many forms of knowledge, which stifles both innovation and technology diffusion.

This issue needs to be looked at afresh if technology transfer has to be meaningful. The trade-related aspects of intellectual property rights (TRIPs) agreement, therefore, should not solely focus on protecting IPRs; its goal should be to create an IPR regime that provides IPR incentives for technological innovation, global diffusion and usage of that innovation. Both industrialized and developing nations share responsibility for overhauling TRIPs and domestic IPR laws to achieve climate change mitigation objectives.[58] This issue has yet to be given sufficient attention in the climate change debate.

REPAIRING ALLIANCES

Brazil, South Africa, India and China, at a meeting in New Delhi, indicated their support for the Copenhagen Accord on climate change. It is understood that these four countries will try to mend fences with those G77 members who tend to believe that the more advanced developing countries (particularly these four) had broken ranks with the G77 in Copenhagen. It is also understood that these countries are considering establishing their own funds to provide assistance, mainly at the bilateral level, to less-developed and vulnerable countries. It is reported that India's initial proposal received a favorable response from South Africa and Brazil, and they are now exploring the idea further with their respective caucuses.[59]

For its part, India has pledged to try and reduce the emission intensity of its GDP by 20–25 percent by 2020, in comparison to the 2005 level through domestic mitigation actions with the caveat that these actions will be entirely voluntary and not legally binding.[60] Also, such mitigation actions will not apply to the country's agricultural sector. India has insisted on preserving the centrality of the UNFCCC process, and necessary steps must be taken that are in line with the discussions at the conference of parties (CoP-15), the two-track negotiating process Ad Hoc Working Group (AWG) on Long-term Cooperative Action and the AWG of the Kyoto Protocol. All need urgent reactivation in the lead-up to CoP-16. Just how many suggestions raised at such forums are agreed to remains unknown at this stage.

The Copenhagen Summit took place in the middle of a financial crisis triggered by the failure of the US banking system. Countries are slowly recovering from this crisis, as their various financial packages start to take effect. However, the recent crisis in EU countries has rattled the EU financial system to its core. Aside from the United States, the countries in the European Union have usually been major providers of financial support to developing countries. Experts believe that the world has not yet seen the end of the financial crisis within these EU member countries, and it will be some time before their markets and banking systems stabilize. Also, Japan's international credit rating has been recently downgraded because of its high foreign debt. Collectively, these countries provide a major portion of the development funding and specific purpose grants assistance to developing countries. There is thus a question as to who will provide the considerable funds required to meet the required restructuring and climate change–mitigating measures. The prognosis for the Cancun summit remains uncertain.

Nonetheless, there has never been a single conference or summit of this kind with more than one hundred of the world's leaders in attendance. It also must be acknowledged that climate change is a multifaceted issue. Consequently, despite a common global objective, Revkin argues that each country has its own priorities and sociopolitical compulsions, "They're individual countries with individual plans and projects, but together they do represent a world that's slowly kind of, after a couple of decades of debate, coming to grips with this looming issue and trying to kind of shift the curve."[61]

He further argues that it took two decades to agree to an outcome that establishes a trajectory that could reach the ultimate objective. He compared the climate change issue with the form of trade negotiation issues in existence since the end of the Second World War in 1945. The *Economist* agrees with the thrust of Revkin's argument, commenting that

> the UN's climate process has for more than a decade been bedevilled by a binary split between developed and developing countries. The Copenhagen accord has achieved some progress toward closing the split suited to the between [*sic*] developed and developing countries; and
>
> Climate change is an unusually complex issue, which crosses and confounds the boundaries that normally define our world. . . . Trying to deal with all the sources of the many gases involved in a single set of negotiations, in a forum of 193 countries, was always a tall order.[62]

Indirectly, the *Economist* also conceded to the importance of the sovereignty issue when it argued that negotiators need to be aware of the reaction of the US Senate to any agreement reached at Copenhagen. The issue is not fundamentally different from the position adopted by China (which faced some disapproval) of not allowing external agencies to measure domestic emission levels, notwithstanding any technical difficulties.

Experience to date has made it clear that rapidly developing countries will not agree to adopt a CO_2 emission reduction target until they reach a certain level of economic development. At the same time, industrialized countries refuse to take unilateral action to reduce legally binding emission targets.

The emphasis, therefore, has to be on encouraging individual countries to act voluntarily through individual initiatives to reduce CO_2 gas emissions by all mitigation instruments available to them. And, at the same time countries should seriously consider (of course in good faith, which oftentimes has been lacking in the past during both the Cold War and post–Cold War period) that while mitigation measures might appear to be costly, at the end of the day (as far as one can ascertain at this stage) a green economy would have more benefits than costs for both for developed and developing economies. Countries need to consider all issues with a genuine intention to make some short-term economic sacrifice; as the old dictum goes *there is no gain without pain*. China's initiatives appear to be working for them (chapter 5) and other countries can learn from its experience. Also, the resolution adopted by the four rapidly industrialized countries (Brazil, South Africa, India, and China) to consider the gas emission issue as an architectural building block appears to be the second-best option available to achieve the ultimate objective.

Last, but not least, the best option is to adopt a legally binding target; but, should the global political leadership fail to act, there is not much any less-developed country can do, given their limited capacity. Nor would any success by such a country mean much, as, in the global context, their achievement would not create a canopy to protect the world from harmful gas emissions into the atmosphere.

NOTES

1. Aldred 2009.
2. Stavins 2009.
3. Rogelj et al. 2010. Also, "Copenhagen Sets Earth for More Warming," http://www.abc.net.au/science/articles/2010/04/22/2880122.htm?topic=enviro (accessed April 27, 2010).
4. Barrett 2009.
5. Marull 2010; "Power and the Xingu," *Economist*, May 7, 2010, http://www.economist.com/research/articlesbysubject/displaystory.cfm?subjectid=348924.
6. McCully 1996; Rich 1994.
7. "Tap That Water," May 7, 2010, http://www.economist.com/research/articlesbysubject/displaystory.cfm?subjectid=348924.
8. Bryson 2009.
9. Lunt et al. 2009.
10. Kamel Djemouai is chairman of the Africa Group.
11. "China's Thing about Numbers," *Economist*, December 30, 2009, http://www.economist.com/world/international/displaystory.cfm?story_id=15179774.
12. Aldred 2009.
13. In an interview with *NewsHour* on December 24, 2009.
14. "Copenhagen Disappointing: PM," *Statesman*, January 4, 2010.

15. "Jairam Bid to Rework Stand on Climate?" *Times of India*, February 22, 2010.

16. R. Puri, "India's Achievement in Copenhagen," *Statesman*, December 23, 2009; "China's Newest Puppet," *Statesman*, December 27, 2009.

17. Salmon Haidar, "Hazy and Fuzzy: Much to Weigh and Assess, Post Copenhagen," *Statesman*, December 24, 2009.

18. Professor Tim Flannery in an interview with ABC's *Lateline* program on December 14, 2009; "Good Climate Agreement Brewing: Climate Council," http://www.abc.net.au/lateline/content/2009/s2771630.htm.

19. Aldred 2009.

20. Weitz 2009.

21. http://www.digitaljournal.com/article/288008 (accessed May 10, 2010).

22. *The Mantle*, http://www.mantlethought.org/content/latin-american-and-caribbean-leaders-new-alli (accessed May 10, 2010).

23. "Rio Group to Create Hemispheric Organization Excluding US and Canada," http://latindispatch.com/2010/02/23/rio-group-to-create-hemispheric-organization-exc (accessed May 10, 2010).

24. Lehmann 2010.

25. Rhein 2010; Mùller-Kraenner and Kreme 2010.

26. CE Delft, an independent Dutch research and consultancy firm, argued that the EU should have aimed for 40 percent carbon dioxide emissions cut to hold back global warming; "EU Should Step Up Climate Cut," *Times of India*, March 16, 2010.

27. CEO climate policy recommendations to G8 leaders, "Executive Summary," July 2008, http://www.weforum.org.

28. "5-Day Davos Forum Ends on Note of Humility," http://news.yahoo.com/s/ap/20100131/ap_on_bi_ge/davos_forum (accessed February 1, 2010).

29. "Bush Backs India for UNSC, but Says It Is Difficult," http://global.factiva.com/hp/printsavews.aspx?ppstype=Article&pp=Print&hc=public (accessed December 31, 2009).

30. Barrett 2009.

31. A. Revkin, "What Next after Copenhagen?" December 18, 2009, http://www.abc.net.au/lateline/content/2009/s2776473.htm.

32. Marion Wilkinson, environment editor of the *Sydney Morning Herald*, indicated a few days before the main conference session that she had "been told by negotiators [in Copenhagen] that there is a fear from the Chinese and the Indians. They fear that the verification measures put in place could be used against them, especially by the US Congress, also perhaps by some of the European Parliamentarians, to impose carbon tariffs on them; that this will be used as a weapon to slug them in the international trade sphere." If anything, this brings to the surface the level of fear and distrust over the issue of carbon tariffs and trade issues; http://www.abc.net.au/7.30/content/2009/s2771588.htm (accessed December 21, 2009). China bridles at this notion, as it is acutely and painfully aware that for a long time it had no control over its economy and trade and the country was sliced up by the Western powers. Its burgeoning economy, its feel-good feeling that it had already made bigger concessions, its confidence on its own ability as a big-time player at

the global stage provided the strength to remain steadfast on its position that it does not need that regime in place.

33. *Lateline*, December 17,2009, http://www.abc.net.au/lateline/content/2009/s2775218.htm.

34. "Climate Boss Admits Copenhagen Failed," http://www.abc.net.au/news/stories/2010/01/21/2797558.htm?site=news (accessed January 23, 2010).

35. "A Look at Climate Change after Copenhagen," *NewsHour*, December 21, 2009.

36. Miliband 2009.

37. 2007 figure; HDR_2009_Tables_rev.xls (accessed February 2, 2010).

38. "Developing Countries Protest Copenhagen Proposal," http://www.abc.net.au/lateline/content/2009/s2767039.htm (accessed December 21, 2009).

39. Mr. Rasmussen, Danish prime minister, publicly acknowledged and commented, "Global warming known no borders. It does not discriminate, it effects us all and we are here today because we are all committed to take action"; http://www.abc.net.au/lateline/content/2008/s2764519.htm (accessed December 12, 2009).

40. Shields and Brooks on climate deal, in an interview with the PBS *NewsHour*'s Jim Lehrer, December 18, 2009.

41. Lehmann 2010.

42. "Blowing in the Wind: Discovering Which Countries Emit Which Greenhouse Gases Is Tricky," *Economist*, December 15, 2009, http://www.economist.com/world/international/displaystory.cfm?story_id=15108365.

43. Clarke 1999; Lean 1999; Bunting 1999.

44. Ray 2003: 124–25.

45. http://www.abc.net.au/lateline/content/2009/s2771625.htm, December 14, 2009.

46. T. Flannery, "Good Climate Agreement Brewing: Climate Council," December 14, 2009, http://www.abc.net.au/lateline/content/2009/s2771630.htm.

47. Stavins 2009.

48. Barrett 2009.

49. Aldred 2009.

50. Barrett 2009.

51. "Short Changing the Planet," BBC, November 25, 2009, http://www.bbc.co.uk/worldservice/documentaries/2009/11/091125_wednwesdaydoc_shortchangingtheplanet.shtml; "Climate Cash Is 'Unaccounted' For," *BBC News*, November 25, 2009, http://newsvote.bbc.co.uk/mpapps/pagetools/print/news.bbc.co.uk/2/hi/business/8376009.

52. "Short Changing the Planet," BBC, November 25, 2009; Copenhagen-related news/documentaries from its archives.

53. A. C. Revkin, "*NY Times* Environment Reporter Joins *Lateline*," 2009 http://www.abc.net.au/lateline/content/2009/s2767053.htm.

54. "Greek Budget Crisis Could Stagnate US Recovery," *NewsHour*, March 8, 2010, http://www.pbs.org/newshour/bb/business/jan-june10/greece2_03-08.html; "Greek Seeks Washington's Support to Curtail Market Speculation," *NewsHour*, March

8, 2010, http://WWW.pbs.org/newshour/bb/business/jan-june10/greece1_03-08.html; and *Washington Post*, March 3, 2010.

55. "Prof. Robert Schiller Joins the 7.30 Report," http://www.abc.net.au/7.30/content/2010/s2834668.htm (accessed March 4, 2010).

56. "New Ideas Needed to Finance Climate Solution: IMF," http://www.abc.net.au/lateline/content/2009/s2774017.htm (accessed December 21, 2009).

57. "IMF Plans 100 Billion Dollar Fund to Help Poor Mitigate Climate Change," *Times of India*, February 2, 2010.

58. Littleton 2009.

59. "Copenhagen & Beyond: Stage Set for BASIC Meet in Delhi," *Times of India*, January 22, 2010, and "Basic Countries Accede to Copenhagen," *Times of India*, January 24, 2010.

60. Mathur 2010.

61. Revkin 2009.

62. Revkin 2009; "Planet B: How the Underwhelming Copenhagen Accord Could Turn into a Useful Document," *Economist*, December 30, 2009, http://www.economist.com/node/15174523/print.

Chapter Eight

A Few Concluding Thoughts

For a variety of historical reasons India finds itself in a region of either ongoing or recently resolved armed conflict—consider Afghanistan, Pakistan, Bangladesh, Burma, Sri Lanka, and Nepal. Internally, India faces security challenges that range from terrorism to militant (Maoist/Naxal) movements. Violence, whether internal or external, has become a way of life for India. One example in mid-2010 was the increased reports of violent incidents by the Maoists, resulting in numerous deaths. The Indian Minister of Home Affairs, P. Chidambaram, acknowledged that the trouble had spread to over twenty states.[1] India's policy response to environmental and water crises has to be within the realm of political and fiscal reality.

The greenhouse gas emission and related issues such as land degradation and salinity, biodiversity, deforestation, and water crisis are matters of serious concern. Although India and Pakistan are often at loggerheads, they have one common task ahead of them in confronting climate change. Issues in Nepal and Bhutan are different. Their primary concern is melting glaciers, soil erosion, and changes in rainfall patterns. Issues in Bangladesh are similar to those of India, except that land salinity is not a major concern, although salt water intrusion in the low-lying areas remains a serious threat.

In the view of industrialized economies, CO_2 gas emission levels in India are a major global concern. These countries pair India with China in one category. In particular, the United States, along with few other industrialized nations, refuses to agree on a legally binding CO_2 gas emission target unless India and China also commit to such targets.

This argument defies logic. First, the current high level of CO_2 emission in the atmosphere is caused by a high level of continuous emissions in industrialized countries; this has been so since the days of the Industrial Revolution. They achieved their economic growth riding on the back of an enormous gas emission into the atmosphere. Countries such as China and India are in their initial stages of economic growth. They cannot afford to reduce their dependence on fossil fuel lest their economic growth be severely hampered. Secondly, and more importantly, per capita CO_2 gas emissions in both these countries are very low compared to industrialized countries.

India argues that the Copenhagen Accord is not to be treated as the basis of a negotiating text unless it is agreed and adopted by all parties and that it should be used to facilitate formal UN talks on the Kyoto Protocol and the long-term agreement. India is asking for a two-track approach: stronger Kyoto Protocol targets for developed countries, and voluntary actions for the developing world.[2] India's stand appears to have been necessitated by fellow developing nations disapproving of India's adopted position at the Copenhagen Summit. India's standing declined both internally and outside India.

It appears that India is keen to restore its lost ground by keeping developing countries' interests in mind. The four recalcitrant countries (according to most of the developing countries at Copenhagen) were India, China, Brazil, and South Africa. China follows its own path to preserve and maintain its relations with other countries, particularly with smaller countries. South Africa was a member of the late-night drafting group, but restored its credentials in the eyes of other developing countries by fiercely criticizing the draft and the way the declaration was adopted. These four countries have decided to recast the debate about climate equity and hold a special meeting in Brazil in June 2010, with the objective to develop a collective position.

Two steps taken by India indicate its intentions to pursue this line vigorously. First, both India and China have adopted a path of bilateral cooperation in the forestry sector, to create a large carbon sink to help absorb greenhouse gases. China is well equipped to lead in this area, as it has already achieved good results. China's aggressive stance is clear from the fact that for the next decade it has set a target of forty million hectares of land under green cover. India would be able to offer only a total of four million hectares in the next ten years.[3]

Second, at the recent meeting in Bonn, India adopted a stronger position than before, which is basically opposite to the US position. The United States wants to disregard the Kyoto Protocol in favor of a new regime, but China and India are totally opposed to any legally binding emission target.

As a prelude to the November meeting in Cancun, Germany and Mexico called the Bonn meeting, and are keen to put the damaged Copenhagen initiatives back on a solid track. The meeting was attended by about forty invited nations and the outgoing UN climate negotiator. They were mindful that the Copenhagen meeting failed to produce the desired outcome, because it was too large and fragmented to achieve any agreement on such a complex issue as climate change. The outgoing UN negotiator argued that it would be impossible to overcome the hurdle at the Cancun meeting. The economic differences between countries are too wide to adopt a cohesive approach that would satisfy all parties.

In Cancun, a battered Kyoto Protocol survived, but progress on mitigation remained elusive because of well-mounted pressures from vested interests that include highly organized resource-rich fossil fuel lobby and some industrialized countries.[4] It, however, provided some hope for two reasons. First, it agreed to the baseline that emissions cuts must be relative to 1990 and not 2005; and second, it included carbon capture and storage technology as a potential offset for rich countries. It allows rich counties to offset their emissions by investing in renewable energy projects in developing countries and by carbon capturing underground. Some remain skeptical because of their bitter experience over the years and argue that developed countries that have met their Kyoto Protocol targets have done so either by accident or by accounting fiat, and those who won't meet their targets either refused to be bound by the protocol or simply repudiated their obligation under it, with no serious consequence.[5] They have also raised doubts as to whether it can be sustained in the long run.

Addressing climate change has financial and economic implications; many believe (ostensibly under pressure from polluting industries) that financial imposts would destroy the competitiveness of their industries. Some industrialized countries even actively considered allocating free or subsidized emissions trading scheme or carbon-trading permits, but could not obtain the necessary legislative approval to implement such policies. A group of researchers (Climate Group) has found that the competitiveness issue was overblown. They reviewed eighty-four corporations, thirty-six city and seventeen regional governments and found that, in aggregate, they emit about 8 percent of global CO_2 emissions, and at the same time have collectively managed to cut their emissions by 14 percent and still make economic gains.[6]

The Bonn meeting agreed that it would be far better if the Cancun meeting were to concentrate on the sticking points of the envisioned treaty, such as cutting greenhouse gas emissions, financial aid from rich to poor countries, technology transfers, or measures to preserve the earth's forests. The meeting dubbed this approach as functioning architecture (or building blocks) that would provide nations with the tools to fight the climate change issue effectively, and eventually increase the level of participation. At the end of the meeting, UN climate chief Yvo De Boer said that he expected the elusive treaty to become a reality, but that even this would not provide a definitive answer to the climate change challenge.[7]

In this context it is worth noting that some had doubts even before the Copenhagen Summit that a gathering of about 192 countries may not reach agreement on complex and disparate issues ranging from energy security to development assistance. To achieve agreement, more local or national interest might have to be added to the mix to bring countries onside long enough to generate enough positive responses to make the climate change package broadly acceptable.[8] The outcome from the Copenhagen Summit confirmed their concerns. It is possible that even this new approach may fail to achieve any legally binding agreement.

The environmental issues in South Asia are dire. Temperature rise, (possibly) declining rainfall, reduced food production, soil degradation, and rapidly declining per capita water availability are all key issues. More importantly, these issues are all interrelated and require significant investment to overcome their consequences.[9] Many of these deficiencies have been with us for a long time, but coordinated and adequately structured efforts have been lacking. This relates directly to the governance issue. For example, even after spending tens of millions of dollars, pollution levels in the highly polluted Ganges River increased to 100 percent in 1999 from 80 percent in 1994 (see chapter 2). India's Supreme Court expressed its displeasure with the misuse of the Ganga Action Plan funds when the government announced another new program.

India's initiative to tap into China's experience with regard to its own forestry sector is a positive move. China has approached the climate change issue directly and with zeal. This has been missing in India. For example, for some time both France and Japan have been processing human excrement primarily for decorative ornamental purposes and for the extraction of protein from the excrement for use in food production.

China is using human excrement for biogas production, initially for cooking gas and domestic lighting purposes. The residue of the excrement-mud is used to make bricks for paving, large-scale worm farming for enriching soil, and as an input in both the food and cosmetic industries. They now believe that they will be able to extract minerals from waste to produce plastic.

Interestingly, China is integrating this industry with pig farming because pig excrement is rich in energy, and mixing pig excrement with human excrement during the bacteria formation stage enhances both the quantity and quality of the final output. China is now intensively researching uses for cow manure. It already has two million biogas digesters and plans to build a further one million digesters this year.[10] Professor Ross Garnaut, a well-known Australian academic, economist and former ambassador to China, was right when he publicly praised China for its total approach to environmental issues.[11] Many may view this approach as symbolic, but symbolism has its own way of generating enthusiasm.

This skillful utilization of a waste product contributes in preserving the quality of land, as people are no longer required to collect wood for cooking purposes. As a consequence, such lands remain available for agricultural production. Many readers will be aware that Haiti was once self-sufficient in food production. Since the devastating earthquake on January 12, 2010, the inhabitants of Haiti continue to suffer severe food shortages and are living on handouts and aid shipments.

Two unfortunate reasons for the tragic situation in Haiti are, first, all available land has virtually become unproductive after years of inhabitants using cover from the land for fuel, and, second, the United States sent subsidized rice to cover the critical food shortage. An unintended consequence of this policy was that the small amount of cereal Haiti used to produce became uneconomic. In April 2010 former US president Bill Clinton publicly regretted his administration's decision. Meanwhile, China formulates its policies with a clear vision of how it can optimize a social-political and economic return on minimal outlay.

Water is intimately connected with the greenhouse gas emission issue. Shortage of freshwater in a country where rainfall occurs only three to four months a year creates severe problems even in the best of times. Water is both an emotive and a survival issue for all citizens. Despite their critical importance, water management issues have remained archaic and outdated in the subcontinent. Because of its inherently complex nature, the issue requires a process of continuous planning, which unfortunately has not happened in the subcontinent.[12]

India, Pakistan and Bangladesh are water-scarce countries and heading toward water stress. Parts of India and Pakistan already face severe water shortages. Bangladesh is short of freshwater when it needs it most. Bhutan and Nepal do not have freshwater crises, but climate change may cause stress over the longer term. Afghanistan has no stake in riparian waters. Its environmental concerns are different to those of other countries in the South Asia region. The water issues need to be considered in the subcontinent from the perspective of the needs of two distinct sectors: municipal (urban) and industrial, and rural.

Realistically speaking, meeting the needs of the urban and industrial sector is relatively easy, because freshwater can be manufactured and made available on a cost recovery basis. Globally, in almost all modern societies, the cost of providing urban water is recovered through water-pricing policies. In India a few urban centers charge for water, but consumers do not directly feel the economic costs of water. Urban citizens are usually well off compared to the nonurban population. There is absolutely no ethical, moral or financial reason for not charging an economic price for providing water in urban areas. Although there are poor people in urban centers who would not be able to pay any price for basic water use, this problem can be overcome by providing free or subsidized water to the really needy section of the population. Success of any such dual policy depends upon a properly developed administrative mechanism, which is missing from the Indian administrative system.

India's existing national water policies are a clear example. Its freshwater policy has been a partisan approach focused solely in favor of a supply-side solution. More importantly, poor institutional capacity, political pressures and corrupt practices created by a bureaucratic-political alliance have contributed to this policy stance. In 1985, an independent study found that irrigation department postings were being auctioned, implying that the power to distribute water brings significant illegal payoffs to various officials.[13] The ground situation has significantly deteriorated since then.

Tentacles of corruption in India have spread all over the society. In 2005, noted journalist C. R. Irani wrote: "Yes, I am leading up to the appalling corruption in the country. . . . While I agree we are not at rock bottom—not yet."[14]

The situation continues to deteriorate. For the second year in a row the Comptroller and Auditor General of India (CAG) declined to take responsibility for the Union finance accounts because government expenditure during 2008–2009 was overstated and lacked transparency. The report specifically mentioned that the government transferred over 830,000 million rupee (about US$16,940 million) directly to state and district level autonomous bodies and nongovernmental organizations for implementation of various flagship schemes, and it was not known how much of these funds remained unspent. Further, the report raised objections over the manner in which the government showed some of its expenditures. The CAG was unequivocal in stating that certain accounting adjustments were not only against the accounting rules; they also resulted in expenditure being incurred without obtaining parliamentary approval.[15]

The corruption environment has not improved since then. High-level government commitment is required, and political leaders must act decisively and transparently to ensure that corrupt practices do not flourish.[16] The situation is worse now: a division bench of India's highest court has ruled in a 2008 judgment that implicated many of India's politicians: "In India, the whole government machinery is corrupt. The government may lay down the law, but who will implement it? In India, this work is done by clerks. The secretaries and joint secretaries are unable to face the clerks. The law is there, the statute is there, but the governments have become non-functional."[17]

As freshwater becomes scarce, the privileged, the powerful, and the well off will use their clout to protect their privileged positions. Any governmental response to the current and future water crisis will remain ineffective, and the judiciary will continue to issue strictures to no avail.

More importantly, a number of case studies conducted by the Pacific Institute in California have found that, in certain cases, long-term sustainable use of water does not require drastic advances in technology or heroic action. Primarily, it requires a commitment to sustainability and efficiency.[18] The University of Adelaide recently completed an extensive interdisciplinary study outlining the issues that must be at the forefront of any water planning and water policy debate. The report received wide recognition, both in Australia and internationally. It found that national and regional water-planning processes require a different policy approach, but that the basic theoretical principles underpinning the report remain fundamentally valid for any water planning exercise.

Three of the report's recommendations regarding water planning and water policy are of particular importance to all riparian countries in the subcontinent. These are

- a reliable, affordable, equitable and healthy supply of water that supports the social and physical environment;
- a community that embraces and contributes to water management in practice and in decision-making processes; and
- policies to provide a more secure and sustainable urban and rural habitats (having learned from historical mistakes).[19]

Lower riparian countries have additional concerns, which must be addressed in their policy context. One of these concerns, from a strategic point of view, is to constantly ensure an uninterrupted flow of unpolluted water from upper riparian countries, including ensuring that adequate procedures are in place to prevent pollution of riparian waterways, particularly through accidental pollution.[20] This is a difficult issue in South Asia, as countries have not shown genuine cooperation in managing riparian waters. Distrust between the countries is so entrenched that even technical information provided by one country becomes unacceptable to others.[21]

For urban centers, freshwater can be produced by recycling or by a desalinization process. In India, recycling water may be a long way off for cultural reasons. Recent research in a nanotechnology-led desalinization method confirms that in the near future a nano system involving carbon tubules will produce 86,000 liters of freshwater per square meter from a membrane, instead of 86 liters per square meter of a traditional membrane. This has already been achieved in a laboratory trial. The process uses just 10 percent the energy of conventional pressurized desalination.[22] In India the quality of urban water is poor and if the total cost of maintaining plumbing systems, and poor quality water induced wear and tear are included, the nano method might eventually prove to be competitive. Technological improvements could cause production costs to fall further over time.

For the rural sector, the task is complex. First, it would be difficult to charge an economic price for the water for food production in poverty-stricken subcontinental countries in South Asia; it is simply not politically feasible. The traditional approach to increase food production has been to replace low-yielding crops with high-yielding crops,[23] improved agricultural infrastructure, and political and technical leadership.[24]

Second, encouraging production of those items that require less water can only be achieved in conjunction with a massive education program to educate people about the nutritional values of alternative types of food. A major constraint in the subcontinent is that rice production requires a large amount of water, much of which is wasted through evaporation.[25] Most root crops require a reduced amount of water per unit of production. Third, encouraging judicious use of fertilizer would lead to reduced water consumption. As these issues are interrelated, they require a coordinated approach.

As the 1997 draft UN Protocol has yet to be ratified by the required number of countries, most riparian water disputes have to be resolved by mutual agreement and, failing that, by arbitration. As arbitration is not value neutral, problems often linger as warring parties ponder their options. This provides opportunities to affected countries and outside unscrupulous elements to play a negative role that may not be always conducive to dispute resolution. Since China has now admitted to constructing a dam on the Brahmaputra River, India should make a concerted effort to develop a long-term strategic plan in conjunction with China and Bangladesh to ensure that all three countries receive a fair share of water from the Brahmaputra. This is important, as China's long-term policy concerning the waters of the Brahmaputra remains vague.

While irrigation facilities have to be extended to bring additional land under irrigation, canal irrigation has to be discouraged for two reasons. First, the evaporation rate is high in South Asia, and this is likely to increase with the predicted temperature rise. Second, the land required for developing canals for irrigation purposes is very high—between 5 percent and 13 percent of newly irrigated land is usually required for reservoirs, canals, and drainage infrastructure.[26] Further, canal irrigation encourages water waste and higher fertilizer consumption. As a consequence, if an appropriate drainage system is not developed along with canal irrigation facilities, land quickly accumulates salt.[27] Developing an appropriate drainage system is expensive and beyond most small farmers' capacity in South Asia. Instead of canal irrigation, farmers should be encouraged to develop well irrigation. Gujarat state, on the border with Pakistan, has achieved success in this area.

In the context of widespread poverty and highly skewed income distribution, the Maoists have succeeded in Nepal and the news of their success is spreading in India.[28] The social unrest created in the process only complicates the situation, as no water policy can satisfy everybody and the rebels can easily use this dissatisfaction to achieve their political objectives. Apart from scattered conflicts that have sprung up in post-independence India, further conflicts also developed after the 1960s, as state-level ethnic politics assumed sharper overtones in places such as Assam, Tamil Nadu, Nagaland, Mizoram, Haryana, the Punjab, and even in highly industrialized Maharastra. As an article in the *Economist* observed, "Their [states'] quarrels are bitter. . . . The southern state of Tamilnadu claims that neighbouring Karnataka violates commitments to share water from the Cauvery. If the Cauvery problem is not solved (wrote one Tamilnadu M.P. this month), Tamilnadu could go the way of Kashmir."[29]

In India, intrastate and interstate water disputes are becoming increasingly bitter and even the Supreme Court finds it difficult to resolve such disputes. Recently the Court asked the state administration in the Punjab and Haryana to release more water for their joint capital (Chandigarh), and the states voiced their reservations about any release of water for their capital.

The rise of Islamic fundamentalism in Bangladesh and Pakistan offers opportunities to those who want to achieve their own political objectives by harnessing discontent. Riparian water provides an opportunity, as its political enemies live outside the bounds of the ordinary citizen. Further, others will willingly use any such opportunity to achieve their narrow political self-interests.[30] Naqvi in the *Far Eastern Economic Review* wrote, "India faces a viscerally hostile neighbour that also happens to be, along with Saudi Arabia, the world's leading exporter of Sunni fervor. . . . Bangladesh appears to be evolving into a somewhat tamer version of its fraternal twin."[31]

Pakistan's problems are complex, as its civilian government is shaky and trapped between Allah, the Army, and the United States. Relations between India and Pakistan are hostage to conflicting public emotions.[32] Pakistan's most pressing challenges reflect the accumulated consequences of decades of clumsy handling of its political, social, and international affairs. In the event the government fails to confront these challenges soon, the consequences could be disastrous for Pakistan,[33] and India may also feel these consequences.

The Pakistan-Afghan conflict has implications for India-Bangladesh détente. Some insurgents coming out of Iraq and Afghanistan infiltrate Indian Kashmir. This is happening with the support of sections of Pakistan's Inter-services Intelligence (ISI) and a section in the Bangladesh administration that allows Dhaka to be a transit for sending jihadi groups to India. A *New York Times* report reveals that anonymous sources with classified information have noted, "[Pakistani officials] said contacts were less threatening than the American officials depicted and were part of a strategy to maintain influence in Afghanistan for the day when American forces would withdraw and leave what they fear could be a power vacuum to be filled by India, Pakistan's arch enemy."[34]

A feeling of hopelessness, a higher incidence of poverty, and widening inequality along with the speed of economic growth[35] certainly complicate the riparian water-sharing arrangements in the subcontinent.

Massive investment is required to address climate change consequences and overcome predicated water shortages. These problems also become a public finance and welfare issue, rather than merely a security and strategic issue. Special policy provisions are needed to ensure that benefits from development policies reach those people who are genuinely disadvantaged. In the past, benefits often went to well-off individuals at the expense of the needy.[36]

Any policy required to face the climate change crisis and the imminent water crisis situation in subcontinental countries boils down to governance issues.

The Indian technocrat-bureaucrat-vested interest alliance is pursuing policies that it considers best for the country. One such policy concerns the hastily designed river-linking projects. The decision to build a number of dams upstream (such as the Bhairon Ghati I, Bhairon Ghati II, Loharinag-Pala, and Pala-Maneri) has raised a number of environmental, ecological, and financial issues. India is an upper riparian country to Bangladesh, and it would not wish to be isolated from any deliberations on water-sharing arrangements with the riparian countries to its north. Failure to provide legitimate answers only raises peoples' suspicion, both domestically and internationally.[37]

Already noted are examples in which India's Supreme Court expressed frustration because its judgments went unheeded. In India people can use public initiated litigations (PIL) for public good, and the court decides whether to accept them. In a democracy, when the administration fails to fulfill its constitutional duties, the PIL provision is a boon that allows the judiciary to come to the rescue of the suffering applicants. One such ruling ordered the government to form a panel of scientists to take up research on a war footing to tackle the country's water problem.[38]

The Indian government has not yet directly responded to the direction of the court, except in October 2009 when Prime Minister Manmohan Singh sought caution on river-linking projects. When the government initially proposed the immense river-linking project, despite not carrying out an environmental impact assessment statement or financial cost-benefit analysis, it hurriedly started implementing stages of this project. A large number of environmentalists, economists, and scientists cautioned the government and objected to this project, but all were overruled.[39] Now the prime minister is saying that the project might cause "irreparable damage to our ecology" and "I believe we have to move with caution."[40] Interestingly, at the political level, the general secretary of the ruling party is alleged to have said "it can lead to ecological disaster," while the environment and forests minister, Jairam Ramesh, is alleged to have said that linking the rivers is not a good idea considering the significant expenditure involved and the potential for environmental damage. The prime minister has also said that "Expert opinion will be taken."[41] It is now clear that the government's water policy is in tatters; yet only a few years ago, against domestic and international objections, it wanted to complete the river-linking project by 2016.

History tells us that industrial societies defend policy decisions based on rational self-interest.[42] This has not been the case in South Asia. The Indian government must weigh all policy alternatives properly, acknowledging that it has failed to learn from past mistakes or from successful governance practices elsewhere.

Given the uncertain political environment, it is difficult to predict how the situation will evolve. India and China, as populous countries, can play a major role. What they do will largely influence future developments.[43] The political direction of the two Islamic countries (Bangladesh and Pakistan) and their respective capacities to promote and develop civil societies are also critical to future developments, and this needs to be a critical focus for India and China when determining their future strategic direction. The political situation in Nepal remains fluid, but other countries must provide it with sufficient space to play a legitimate role in the region's (and the basin's) environmental and water policies.

One irony of the postcolonial development process is that various parties (whether individual donors, the World Bank, or a UN agency) place insufficient emphasis on using local knowledge when determining their policy priorities. This has contributed to community indifference and apathy, which always contribute to policy failure and poor governance practices. The importance of grassroot participation in policy development cannot be overemphasized, and the farming sector is no exception. Development experts are even considering abandoning the notion of "donor knows best."[44]

Last, but not least, South Asian countries must recognize that cooperation is a sensible approach to resolving crises and preventing conflict, although these mostly have remained outside their policy domain.[45] Not doing so has prevented them from optimizing their development and growth potential, and in the words of veteran environmental journalist Andrew Revkin, "I have learned a lot about . . . climate science and the robustness of this overall picture. But there's clearly a dynamic there, where they are so frustrated that they have not been able to get society to understand the risks of long-lived gas accumulating in ways that are hard to reverse."[46]

Just as all matter has evolved down the centuries, so the climate is also constantly changing. We experience some of these changes, but most occur silently. Almost every aspect of our surroundings is influenced and affected by the changing climate, and these changes, in turn, also continuously induce further change around us.

Climate change usually refers to the phenomenon known as global warming or anthropogenic global warming. Most physical changes that have contributed heavily to climate change occur continuously. These elements are tectonic plate movements, variation in solar output, volcanic eruptions, alterations in the earth's orbit and ocean temperature, and last, but not least, human-induced changes.

Of all these elements, current concern is focused on human-induced factors; however any debate around the impact of human activity on the earth's changing climate remains a contentious issue. Globally, knowledge concerning the creation of the universe, its unique astrophysical characteristics, and celestial mechanics are far from accurate or complete. Despite such failings, our knowledge of the universe has improved exponentially in recent decades.

Scientists are uncovering the mysteries of nature, and these discoveries are helping us to better understand the climate change phenomenon. For example, some Chinese scientists now consider that the problem of geological instability and frequent landslides are problematic enough to make any huge hydraulic projects difficult and even impossible.[47] The National Aeronautics and Space Administration found that large volumes of water stored by dams have shifted the angle of the Earth's axis by approximately 60 cm since 1950.[48] No doubt such large masses of stored water also have ecological, climatic, and health consequences. The global community and the political leadership can only ignore these warning signs at their peril!

NOTES

1. Bubna and Shukla 2010.
2. "India Toughens Stand in Runup to Mexico," *Times of India*, May 7, 2010.
3. "India, China Begin Cooperation to Tackle Global Warming," *Times of India*, May 5, 2010.
4. Hewett 2010; Hendrie 2010.
5. Green and Picker 2010; Macey 2010.
6. Motaal 2009.

7. "'Ice Broken' at Climate Meet, but Progress Glacial," *Times of India*, May 5, 2010; and "No Climate Deal This Year, Says UN Negotiator," *Times of India*, May 5, 2010.

8. Heap, Carin, and Smith 2009.

9. "Global Food Prices Rising to Dangerous Levels, World Bank Says," http://www.sofiaecho.com/2011/02/15/1044563_global-food-prices-rising -to-dangerous-l (accessed February 22, 2011); "Global Food Prices: UN Officials, G20 in Calls for Investments in Agriculture," http://www.sofiaecho.com/2011/02/20/1047102_global-food-prices-un-officials-g20-in-c (accessed February 22, 2011); Timmer 2011; "TEXT-G20 finance ministers Paris Feb. 19 communique" http://www.reuters.com/assets/print?aid=USLDE7110FN20110219 (accessed February 21, 2011); Dapice 2011.

10. I viewed an amazing 1-hour documentary. It shows how China has approached the climate change and CO2 emission reduction issues. Its approach includes a wide range of areas. The documentary is titled *The Fabulous Story of Poop: In the Name of Roses*. Multicountry research programs have contributed in making this documentary: Tokyo Bureau of Sewage; Biogas in China; Washington State University; Kosi Fujii in Japan and researcher from France that includes commercial interest as well. The Tokyo Bureau estimates that the output from the human waste could ultimately meet about 60 percent energy needs of households.

11. Professor Ross Garnaut, Australian National University and Melbourne University.

12. Bredehoeft 1997: 35, 41.

13. The study aimed to determine on what basis transfer of officials takes place in the irrigation department in an Indian state. It found that fixed-rate payments were required to get posted to some workstations. Rates of payment to influence posting decisions vary, depending on the level of position. The study found that both local and higher-level politicians were beneficiaries of this arrangement. The researcher recorded an interview where a chief engineer's assistant on personal administration matters (transferring authority) was alleged to have said, "Frankly we are subjected to a lot of pressures from politicians. Politicians are scum of their country, they are dragging it down." The study also found that any department that had the power to inflict costly fines or inflict financial loss by their actions could raise money and therefore command a price for posting in certain high-income areas (see Wade 1985: 467–97).

14. Irani 2005.

15. "CAG Refuses to Take Responsibility for Union Finance Accounts," *Times of India*, May 8, 2010.

16. Ray 2006: 33.

17. "In India, Even God Is Helpless," *Times of India*, August 6, 2008.

18. C. Smith, "California Farm Water Success Stories," *Online Update*, Pacific Institute, March 2010.

19. C. B. Daniels, ed., *Adelaide: Water of a City* (Kent Town, South Australia: Wakefield Press, 2010), 562.

20. See chapter 4. Friedkin 1987. Also, see chapter 2 in this volume.

21. Munir-uz-Zamam 1987.

22. Ellis 2010.

23. Evans 1998: 197.

24. Peterson and Davies 2005.

25. The water required to produce 1 kilogram of rice is, on average, 1,900 liters, compared to 1,650 liters for soybean, 1,100 liters for sorghum, 900 liters for wheat, and 500 liters for potato.

26. World Bank, *India Irrigation Sector Review* (Washington, DC: World Bank, 1991), 7 and 41.

27. Chapman 2000: 109; Chapman and Baker 1992: 30–31 and 61–62.

28. Pai 2007: 10–12; "The Bothersome Little People Next Door," *Economist*, November 6–12, 2004.

29. "Water in India: Nor Any Drop to Drink," *Economist*, August 24, 2000: 49–50.

30. Drezner 2007: 44.

31. Naqvi 2008: 7.

32. Naqvi 2008: 12.

33. Malley 2007.

34. Mazzetti and Schmitt 2009.

35. Wallerstein 2008; Milanovic 2008.

36. Altman 2006; Mollinga, Doraiswamy, and Engbersen 2004; Freeze 1997: 173; and Newberg 2006.

37. See Ganga Bandhan, "The Disappearing River: Facts and Information 2008," May 31, 2008, meetvak@hotmail.com.

38. Singh 2009.

39. Ray 2008a: 115–24.

40. "PM Cautious about River Linking Projects," *Hindustan Times* (Document HNTM 000020091030e5au000gw), htsyndication@hindustantimes.com (accessed January 6, 2010).

41. "PM Cautious about River Linking Projects" (January 6, 2010).

42. Chrislip and Larson 1994: 125–26.

43. Varshney 2007.

44. Werker 2008.

45. Allan 1998. Cited in Chakraborty 2004.

46. *Lateline*, http://www.abc.net.au/lateline/content/2009/s2767053.htm (accessed December 21, 2009).

47. Chen and Chen, "Letter to the Editor," *Nature* 429, no. 6991 (2004): 501.

48. *Harper's Index* 4 (1997): 13, 87.

Bibliography

Adam, D. 2007. "Climate Change: Scientists Warn It May Be Too Late to Save the Ice Caps." *Guardian*, February 20.

Adelaide University. 2010. "New Study Ranks Countries on Environmental Impact." *Science Daily*. Accessed May 8. http://www.sciencedaily.com/releases/2010/05/100505092006.htm.

Agrawal, P., and P. V. Srinivasan. 2000. "Social Development in India." In Tang, *Social Development in Asia*.

Ahamed, E. 1998. "The Military and Democracy in Bangladesh." In *The Military and Democracy in Asia and the Pacific*, edited by R. J. May and V. V. Selochan. Canberra, Australia: ANU E Press.

Akbar, M. J. 2010. "Save Earth from Human Nature." *Times of India*, January 24.

Akhtar, A. S. 2007. "Balochistan versus Pakistan." *Economic and Political Weekly* 42 (45–46).

Alavi, H. 1983. "Class and State." In *Pakistan: The Root of Dictatorship; the Political Economy of a Praetorian State*, edited by H. N. Gardezi and J. Rashid. London: Zed Press.

Aldred, T. 2009. "We Must Not Accept Copenhagen's Failings." Accessed December 30. http://news.bbc.co.uk/2/hi/science/nature/8432927.stm.

Ali, C. R. 1942. *What Does the Pakistan National Movement Stands For?* Islamabad: Pakistan National Commission on Historical and Cultural Research.

———. 1949. *Muslim Minority in India and the Dinian Mission to the UNO*. Lahore, Pakistan: All-Dinia Milli Liberation Movement.

Ali, T. 1983. *Can Pakistan Survive? The Death of a State*. London: Penguin Books.

Allan, J. A. 1999. "Water in International Systems: A Risk Society Analysis of Regional Problem Sheds and Global Hydrologies." Water Issues Group Occasional Paper 22. School of Oriental and African Studies, University of London.

Allan, Tony. 1998. "War and Water." Presented at International Committee of the Red Cross forum. Cited in Chakraborty, "Sharing of River Waters."

Altman, D. 2006. "Managing Globalization: The Trouble with Water." *International Herald Tribune*, November 17.

Anthony, A. 2007. "Musharraf Pressured as Lawyers Protest." Accessed May 11. http://news.yahoo.com/s/nm/20071104/ts_nm/pakistan_dc&printer=1;_ylt=AkaPBKpF.

Arthur, O. 2001. "Opening Statement at the Small State in the Global Economy Conference." In *Small States in the Global Economy*, edited by D. Peretz, R. Faruqi, and E. Kisanga. Economic Paper 44. Sterling, VA: Stylus.

Asian Development Bank. 2006. *Asia Economic Monitor 2006*. Manila, Philippines: Asian Development Bank.

Austin, C. 2004. *Water, Wit and Wisdom: The Search for Solution to the Water Crisis*. Lilydale, Australia: WaterRight.

Australian Bureau of Agricultural and Resource Economics. 2007a. "Climate Change Impacts on Australian Agriculture." Accessed December 15. http://www.abareeconomics.com/publications_html/news/news.html.

———. 2007b. "India's Economic Prospects and Implications for Australia's Commodity Exports." Accessed December 15. http://www.abareeconomics.com/publications_html/news/news.html.

Baker, A. 2007. "Truth about Talibanism." *Time*, March 27.

Bakshi, J. 2004. *Russia-China Relations: Relevance for India*. New Delhi: Shipra.

Banerjee, A. K. 1998. "Rethinking National Security: Domestic and External Sources of Threats to India's Security." In *Security Issues in South Asia: Domestic and External Sources of Threat to Security*, edited by A. K. Banerjee. Kolkata, India: Minerva.

Barber, M., and G. Ryder, eds. 1993. *Damming the Three Gorges*. 2nd ed. London: Probe International and Earthscan.

Barbier, E. B. 2005. *Natural Resources and Economic Development*. Cambridge, UK: Cambridge University Press.

Barlow, Maude. 2007. *Blue Covenant: The Global Water Crisis and the Coming Battle for the Right to Water*. Melbourne, Australia: Black.

Barlow, M., and Tony Clarke. 2002. *Blue Gold*. New York: New Press.

Barrett, S. 2008. "At a Round Table Discussion with Experts (Live on Air) on *El Nino* Issues That Included amongst Others Climate Change; and Policy Implications of Greenhouse Warming." April 6. http://www.abc.net.au/rn/backgroundbriefing/stories/2009/2444935.htm.

———. 2009. "Copenhagen: Yet Another Giant Beginning with an Uncertain End." December 21. http://yaleglobal.yale.edu/content/copenhagen-yet-another-giant-beginning-uncertain-end.

Bate, R. 2006. *All the Water in the World*. St. Leonards, Australia: Centre for Independent Studies Limited.

Bayly, C. A. 1997. "Modern Indian Historiography." In *Companion to Historiography*, edited by M. Bentley. New York: Routledge.

Bayne, N. 1994. "International Economic Relations after the Cold-War." *Government and Opposition* 29: 3–21. Cited in Sperling, "Introduction."

Bell, J. 1994. "Hydrodollars in the Himalayas." *Ecologist* 24 (3).

Bellman, E. 2004. "India Waters, Its Grassroots." *Wall Street Journal*, August 23.

Beresford-Jones, D. 2009. "Forest Clearances Sealed Ancient Civilisation's Downfall." *University of Cambridge News and Events*, November 2.

Bernstein, R., and R. H. Munro. 1997. "China I: The Coming Conflict with America." *Foreign Affairs* 76: 18–32.

Bhalla, N. 2007. "Melting Glaciers, Sinking Isles: Warming Hits India." Reuters News. February 2. http://global.factiva.com/ha/default.aspx.

Bhutto, Z. A. 1969. *Myth of Independence*. Oxford, UK: Oxford University Press.

Billon, P. L. 2005. *Fuelling War: Natural Resources and Armed Conflict*. London: Institute for Strategic Studies.

Black, Maggie, and Janet King. 2009. *The Atlas of Water*. London: Earthscan.

Blatter, J., H. Ingram, and S. L. Levesque. 2001. "Expanding Perspectives on Transboundary Waters." In *Reflection on Water—New Approaches to Transboundary Conflicts and Cooperation*, edited by J. Blatter and H. Ingram. Cambridge, MA: MIT Press.

Blinkhorn, T. A., and W. T. Smith. 1995. "India's Narmada: River of hope." In *Towards Sustainable Development? Struggling over India's Narmada River*, edited by W. F. Fisher. Armonk, NY: M. E. Sharpe.

Blomquist, W., and H. M. Ingram. 2003. "Boundaries Seen and Unseen: Resolving Transboundary Groundwater Problem." *Water International* 28 (2).

Borenstein, S. 2007. "Warming 'Likely' Man-Made Unstoppable." February 2. http://news.yahoo.com/s/ap/20070202/ap_on_sc/france_climate_change,

———. 2009. "The Push for 350: Contradictions and Carbon Levels." December 13. http://news.yahoo.com/s/ap/20091213/ap_on_sc/climate.

Borole, D. V., S. K. Gupta, S. Krishnaswami, P. S. Datta, and B. I. Desai. 1978. "Uranium Isotopic Investigations and Radiocarbon Measurements of River-Groundwater Systems, Sabarmati Basin; Gujrat." In *Isotope Hydrology*. Vol. 1. Vienna: International Atomic Energy Agency.

Bowring, P. 2009. "Bangladesh's Challenge." *Asia Sentinel*, February 20. http://yaleglobal.yale.edu/article.print?id=11997.

Boyle, A. 1996. "Role of International Human Rights Law in the Protection of the Environment." In *Human Rights: Approaches to Environmental Protection*, edited by A. E. Boyle and M. R. Anderson. Oxford, UK: Clarendon Press.

Bradnock, R. W. 1992. "Changing Geography of the States of the South Asian Periphery." In Chapman and Baker, *The Changing Geography of Asia*.

Bradshaw, C. J. A., X. Giam, and N. S. Sodhi. 2010. "New Study Ranks Australia among Top 10 Worst Environmental Offenders." Accessed May 7. http://www.abc.net.au/ra/asiapac/stories/201005/s2892522.htm.

Bradsher, K. 2008. "The Rice Crisis." *Reader's Digest*, August.

———. 2009. "China Tightens Grip on Rare Minerals." *New York Times*, August 31.

Brammer, H. 1990a. "Floods in Bangladesh I." *Geographical Journal* 156 (1).

———. 1990b. "Floods in Bangladesh 2." *Geographical Journal* 156 (2).

Brandt, L., and T. G. Rawski, eds. 2008. *China's Great Economic Transformation*. Cambridge, UK: Cambridge University Press.

Bray, J., L. Lunde, and S. S. Mansoob Murshed. 2003. "Economic Drivers of the Maoist Insurgency." In *The Political Economy of Armed Conflict: Beyond Greed and Grievance*, edited by K. Ballentine and J. Sherman. Boulder, CO: Rienner.

Bredehoeft, J. D. 1997. "Water Management in the United States—A Democratic Process (Who Are the Managers?)." In *Reflection on Hydrology*, edited by N. Burns. Washington, DC: American Geophysical Union.

Brichieri-Colombi, S. 2009. *The World Water Crisis*. London: I. B. Tauris.

Briscoe, J., and R. P. S. Malik. 2006. *India's Water Economy: Bracing for a Turbulent Future*. Washington, DC: World Bank.

Brown, L. R. 2004. *Outgrowing the Earth: The Food Security Challenge in the Age of Falling Water Tables*. New York: Norton.

Browne, P. 2010. "The China Paradox." *Canberra Times*, January 16.

Brummit, C. 2009. "Official: Pakistan to Restore Chief Justice." Accessed March 16. http://news.yahoo.com/s/ap/20090315/ap_on_re_as_pakistan.

Bryson, D. 2009. "South Africa Blasts Copenhagen Failure." December 22. http://www.google.com/hostednews/ap/article/ALeqM5hshRkwauciAwG_LbxRGBwzE7.

Bubna, M., and R. Shukla. 2010. "India's Internal Security Dilemma." *World Politics Review*, January 14. http://www.worldpoliticsreview.com/article.aspx?i4947.

Buckley, C. 2007. "China in Deep Water as Supply Diminishes." *Canberra Times*, December 15.

Bunting, M. 1999. "No End to Malign US Dominance." *Guardian*. Reprinted in the *Canberra Times*, August 25.

Burke, J. 2002. "Land and Water Systems: Managing the Hydrological Risk." In *Water Resources and Sustainable Development*, edited by K. Prasad. Los Altos, CA: Indo American Books.

Buzan, B. 2004. "How and Whom Does China Matter?" In *Does China Matter? A Reassessment*, edited by B. Buzan and R. Foot. New York: Routledge.

Calvocoressi, P. 2009. *World Politics since 1945*. 9th ed. Harlow, UK: Pearson Longman.

Casey, M. 2007. "Agreement Reached on Climate Report." May 3. http://news.yahoo.com/s/ap/20070503/ap_on_re_as/climate_report&printer=1;_ylt=At.

Caspary, G. 2008. "China Eyes Latin American Commodities." http://yaleglobal.yale.edu/displsy.article?id=10212.

Chakraborty, R. 2004. "Sharing of River Waters among India and Its Neighbours in the 21st Century: War or Peace?" *Water International* 29 (2).

Chamie, J. 2004. "Coping with World Population Boom and Burst—Part I. Yale Centre for Study of Globalization." Accessed November 12. http://yaleglobal.yale.edu/display.article?id=4389.

———. 2009. "Population: Some Boom, Some Decline." Accessed March 21. http://yaleglobal.yale.edu/display.article?id=12121.

Chanda, N. 2007a. *Bound Together*. New Haven, CT: Yale University Press.

———. 2007b. "The Double Edge of Globalization." http://yaleglobal.yale.edu/display.article?id=9366.

———. 2009. "Talks of Border Just Fogs the Issue." *Strait Times* (Singapore), May 26. http://global.factiva.com/aa/default.aspx?pp=Print&hc=Publicatio. Document STIMES0020090525e55q0002y.

Chapman, G. P. 1992. "Changes in the South Asian Core: Patterns of Growth and Stagnation in India." In Chapman and Baker, *The Changing Geography of Asia.*

———. 1995a. "Environmental Myth as International Politics: The Problem of the Bengal Delta." In Chapman and Baker, *The Changing Geography of Asia.*

———. 1995b. "The Ganges and Brahmaputra Basins." In Chapman and Thompson, *Water and the Quest for Sustainable Development.*

———. 2000. *The Geopolitics of South Asia: From Early Empires to India, Pakistan and in Bangladesh*. Aldershot, UK: Ashgate.

Chapman, G. P., and K. M. Baker, eds. 1992. *The Changing Geography of Asia*. London: Routledge.

Chapman, G. P., and M. Thompson, eds. 1995. *Water and the Quest for Sustainable Development in the Ganges Valley*. New York: Maunsell.

Chari, P. R. 2003. "Refugees, Migrants and Internally Displaced Persons in South Asia: An Overview." In *Missing Boundaries: Refugees, Migrants, Stateless and Internally Displaced Persons in South Asia*, edited by P. R. Chari, M. Joseph, and S. Chandran. New Delhi: Institute of Peace and Conflict Studies.

Chatterjee, B. 2010. "Headless and Clueless." *Statesman*, January 14.

Chellaney, B. 2004. "Drawing the Line with China." *Japan Times*, July 30. http://www.japantimes.com/cgi-bin/geted.p15?eo20040730bc.htm.

———. 2005. "India-China Rivalry Sharpens." *Japan Times*, April 2. http://www.japantimes.com/cgi-bin/makeprfy.p15?eo20050402bc.htm.

———. 2007. "China Aims for Bigger Share of South Asia's Water Lifeline." *Japan Times*, July 2.

Chey, J. 2007. "The Gentle Dragon." http://yaleglobal.yale.edu/display.article?id=10034.

Chin, G. T. 2008. "China's Evolving G8 Engagement." In *Emerging Power in Global Governance*, edited by A. F. Cooper and A. Antkiewicz. Waterloo, ON: Wilfrid Laurier University Press.

Ching, F. 2009. "Challenges for China at Sixty—Part II." *Yale Global*, September 28.

Chorus, I., and J. Bertram, eds. 1999. *Toxic Cyanobacteria in Water*. London: F & FN Spon.

Chrislip, D. D., and C. E. Larson. 1994. *Collaborative Leadership: How Citizens and Civic Leaders Can Make a Difference*. San Francisco: Jossey-Bass.

Chung, Chien-peng. 2004. *Domestic Politics, International Bargaining and China's Territorial Disputes*. London: RoutledgeCurzon.

Clad, J. C. 1996. "South Asia: Buoyant Economics, Nuclear Weapons, and Environmental Stress." In *U.S. Foreign and Strategic Policy in the Post–Cold War Era*, edited by H. J. Warda. Westport, CT: Greenwood Press.

Clarke, H. 1999. "Poor Nations Are Sidelined." *Canberra Times*, August 7.

Clarke, R., and J. King. 2006. *The Atlas of Water: Mapping the World's Most Critical Resource*. London: Earthscan.

Cline, W. 2007. "Global Warming and Agriculture: New Country Estimates Show Developing Countries Face Declines in Agricultural Productivity." Center for Global Development and Peterson Institute for International Economics. Accessed January 28. http://www.cgdev.org/content/publications/detail/14090#Chpt.

CNA. 2007. *National Security and the Threat of Climate Change*. Alexandria, VA: CNA.

Cockburn, A. 2010. "Sinking Sundarbans." *Statesman*, January 14.

Cody, E. 2005. "China's Quiet Rise Casts Wide Shadow." *Washington Post*, February 26.

Cohen, I. W. 2007. "Chinese Lessons: Nixon, Mao, and the Course of US—Chinese Relations." *Foreign Affairs* 86 (2).

Commission on Human Rights. 2003. *Human Security Now: Protecting and Empowering People*. Edited by S. Ogata and A. Sen. New York: Commission on Human Security.

Conniff, R. 2007. "Third World to Bear Brunt of Global Warming." *Environment Yale*, spring.

Connor, S. 2008. "Scientists Identify 'Tipping Points of Climate Change.'" *Independent* (UK), February 5.

Cooley, H. 2006. "Floods and Droughts." In *The World's Water*, edited by P. H. Gleick et al. Washington, DC: Island Press.

Costanza, R., R. d'Arge, R. de Groot, et al. 1997. "The Value of the World's Ecosystem Services and Natural Capita." *Nature* 387: 353–60.

Crawford, B. 1994. "The New Security Dilemma under International Economic Interdependence." *Millennium: Journal of International Studies* 23: 25–55. Cited in Sperling, "Introduction," and Moran, "An Economic Agenda."

Crow, B. 1995. *Sharing the Ganges: The Politics of Technology of River Development*. New Delhi: Sage.

Dahal, D. R. 2001. *Civil society in Nepal: Opening the ground for questions*. Kathmandu: Centre for Development and Governance.

Daniels, C. B. 2010. *Adelaide: Water of a City*. Kent Town, Australia: Wakefield Press.

Dapice, D. 2009. "Mounting US Debt Burden Threatens Poorer Nations." Accessed March 9. http://yaleglobal.yale.edu/display.article?id=12040.

———. 2011. "The Looming Food Crisis." *Yale Global*, February 18.

Darwin, R. 2001. "Issues in Food Security: Climate Change and Food Security." Agriculture Information Bulletin 765-8. Washington DC: US Department of Agriculture.

Dasgupta, C. 2007. "Climate-Change Challenge for the Poor—Part I." http://yaleglobal.yale.edu/display.article?id=9720.

Dasgupta, S. 2010. "China Mulls Setting Up Military Base in Pakistan." *Times of India*, January 29.

Dasgupta, S., B. Leplante, C. Melsner, D. Wheeler and J. Yan. 2007. "The Impact of Sea Level in Developing Countries." Policy Research Working Paper 4136. Washington, DC: World Bank.

David, D. L., T. Kjellstrom, R. Slooff, and A. McGartland. 1997. "Short-Term Improvements in Public Health from Global-Climate Policies on Fossil-Fuel Combustion—An Interim Report." *Lancet*, November 8.

Debroy, B. 2009. "India's Vicious Downward Cycle." *Far Eastern Economic Review*, March.

de Jonquieres, G. 2004. "Garment Industry Faces a Global Shake-Up." *Financial Times*, July 19.

Dellapenna, J. W. 1999. "Custom-Built Solutions for International Disputes." Accessed March 10, 2004. http://www.unesco.org/courier/1999_02/uk/dossier/txt41.htm.

Dellink, R., M. den Elzen, H. Aiking, et al. 2009. "Sharing the Burden of Financing Adaptation to Climate Change." *Global Environmental Change* 19: 411–21.

Department of Foreign Affairs and Trade. 2008. *High Food Prices, Food Security and the International Trading Systems*. Canberra, Australia: Department of Foreign Affairs and Trade.

Dhakal, S. 2003. "Global Warming Threatens Lake Bursts in Nepal." *One World South Asia*. http://story.news.yahoo.com/news?tmpl=story&cid=655&ncid=655&e=3&u=/onewor.

Dhume, S. 2009a. "Bickering by Champions of Globalization in Asia Worries the Region." *Yale Global*, December 17.

———. 2009b. Review of *Partisans of Allah: Jihad in South Asia*, by Ayesha Jalal. *Far Eastern Economic Review*, January/February.

Diamond, J. 2005. *Collapse: How Societies Choose to Fail or Succeed*. New York: Viking.

Dingle, S. 2009. "Obama in Asia—Part III." *YaleGlobal*. http://yaleglobal.yale.edu.

Dobson, A. 2009. "Malaria and Climate Change/Fighting AIDS, Tuberculosis and Malaria in the Developing World" (includes an interview with Dr. N. Swan). ABC Radio National (Australia). March 2. http://www.abc.net.au/rn/healthreport/stories/2009/2502878.htm.

Dorgi, J., and A. K. Pradhan. 1987. "Water Resources Development in Bhutan." In *Water Resources Policy for Asia*, edited by M. Ali, G. E. Radosevich, and A. A. Khan. Rotterdam, Netherlands: A. A. Balkema.

Drezner, D. W. 2007. "The New World Order." *Foreign Affairs* 86 (2).

Dubash, N. K. 2002. *Tubewell Capitalism: Groundwater Development and Agrarian Change in Gujrat*. Oxford, UK: Oxford University Press.

Dupont, A. 2000. "Food, Water and Security: What Are the Connections?" In *Food, Water and Security*. Canberra, Australia: Australian Centre for International Agriculture Research.

Dupont, A., and G. Pearman. 2006. *Heating Up the Planet: Climate Change and Security*. Sydney, Australia: Lowy Institute.

Durning, A. 1991. "Asking How Much Is Enough." In *State of the World 1991: A Worldwatch Institute Report on Progress toward a Sustainable Society—Worldwatch Institute*. North Sydney, Australia: Allen & Unwin.

Dutta, S. 1998. "China's Emerging Power and Military Role: Implications for South Asia." In *In China's Shadow: Regional Perspectives on Chinese Foreign Policy and Military Development*, edited by J. D. Pollack and R. H. Yang. Santa Monica, CA: RAND.

Dyer, G. 2008. *Climate Wars*. Carlton North, Australia: Scribe.

Dyson, T., T. R. Cassen, and L. Visaria, eds. 2004. *Twenty-First Century India: Population, Economy, Human Development and the Environment*. Oxford, UK: Oxford University Press.

Economic and Social Commission for Asia and the Pacific. 1992. *State of the Environment in Asia and the Pacific 1990*. Bangkok: Economic and Social Commission for Asia and the Pacific.

Economy, E. 2007. "China's Missile Message." *Washington Post*, January 26.

Economy, E., and K. Liberthal. 2007. "Scorched Earth: Will Environmental Risks in China Overwhelm Its Opportunities?" *Harvard Business Review* 85 (6): 88–96. Cited in Friedman, *Hot, Flat and Crowded*, 352.

Ekman, B. 2010. "Roots of Copenhagen Failure: Nature Does Not Recognize Nations." Accessed April 12. http://yalegglobal.yale.edu/content/roots-copenhagen-failure-nature-does-notrecognize-na.

Ellis, A. 2010. "Desalinisation Using Carbon Nanotubes." Accessed April 28. http://www.abc.net.au/rn/scienceshow/stories/2010/2881456.htm.

Estes, R. J. 1998. "Trends in World Social Development: 1970–95; Development: Challenges for a New Century." *International Studies in Sociology and Social Anthropology* 71.

Evans, L. T. 1998. *Feeding the Ten Billion: Plants and Population Growth*. Cambridge, UK: Cambridge University Press.

Fair, C. 2007. "Bangladesh's Fragile Democracy." March 16. http://australiannetwork.com/cgibin/common/printfriendly.pl?http://www.abc.net.au/a.

Fair, C. C., and S. Ganguly. 2007. "Bangladesh on the Brink." *Wall Street Journal*, February 7.

Far Eastern Economic Review. 1988. "Stemming the Flood." October 13.

Faris, S. 2009. "For Which It Stands: Environment." *Globalpost*, January 8. http://globalpost.com/dispatch/environment/090105/which-it-stands-environment.

Felipe, J., and J. Miranda. 2009. "A Plea to Pakistan: Fix Your Economy." *Far Eastern Economic Review* 172 (1).

Ferdinand, F. 2002. "China and Central Asia." In *China's Place in Global Geopolitics*, edited by K. E. Brodsgaard and B. Heurlin. New York: Routledge.

Fishbone, A., ed. 2007. *The Struggle for Water*. London: International Debate Education Association.

Frederiksen, H. D., J. Berkoff, and W. Barber. 1993. *Water Resource Management in Asia*. Vol. 1. World Bank Technical Paper 212. Washington, DC: World Bank.

Freeze, R. A. 1997. "Groundwater Contamination: Technical Analysis and Social Decision Making." In *Reflection on Hydrology: Science and Practice*, edited by N. Burns. Washington, DC: American Geophysical Union.

Friedkin, J. F. 1987. "International Water Treaties: United States and Mexico." In *Water Resources Policy for Asia*, edited by M. Ali, G. E. Radosevich, and A. A. Khan. Rotterdam, Netherlands: A. A. Balkema.

Friedman, T. L. 2008. *Hot, Flat, and Crowded*. London: Allen Lane.

Fukamachi, Y. 2010. “Massive Southern Ocean Current Discovered.” *Nature Geoscience*, April 26. http://www.csiro.au/news/Southern-Ocean-current-discovered.html.

Fullilove, M. 2008. “Worldwide Webs: Diasporas and the International System.” Lowy Institute for International Policy. Accessed February 8. http://www.lowyinstitute.org/NewsRoomItem.asp?pid=1764.

Gall, C. 2007. “Local Militants in Pakistan Add to Quaeda Threat.” *New York Times*, December 30.

Galtung, J. 1985. *Development Theory: Notes for an Alternative Approach*. Berlin: International Institute for Environment and Society.

Gancheng, Z. 2005. “China: Periphery and Strategy.” In *China and Asia's Security: China in the Globalized World Series*. Vol. 1. Shanghai Institute for International Studies. Singapore: Marshall Cavendish Academic International.

Gellner, E. 1999. “The Coming of Nationalism, and Its Interpretation: The Myth of Nation and Class.” In *The Politics and Economics of Power*, edited by S. Bowles, M. Franzini, and U. Pagano. New York: Routledge.

Ghassemi, F., A. J. Jakeman, and H. A. Nix. 1995. *Salinisation of Land and Water Resources: Human Causes, Extent, Management and Case Studies*. Sydney: University of New South Wales Press.

Ghosh, A. 2010. “Civil Society Groups Condemn Proposed Action against Arundhati Roy.” *Times of India*, April 24.

Giggacher, J. 2009. “The Dark Gleam: Gritty Realities on the Indonesian Oil Palm Frontier.” *ANU Reporter* (Australian National University, Canberra).

Gijsberts, M., L. Hagendoorn, and P. Scheepers, eds. 2003. *Nationalism and Exclusion of Migrants—Cross-national Comparisons*. Aldershot, UK: Ashgate.

Giles, J. 2007. “Spend Now, Reap the Rewards Late.” *Weekly New Scientist*, December 1.

Gilmartin, D. 1994. “Scientific Empire and Imperial Science: Colonialism and the Indus Basin.” *Journal of Asian Studies* 53 (4).

Glaser, Boninie S. 2007. “In China, Talk of Democracy Is Simply That.” *New York Times*, April 20.

Gleick, P. H. 1990. *Global Climate Change and International Security*. Oakland, CA: Pacific Institute.

———. 1992. *Water and Conflict*. Occasional Paper 1. Project on Environmental Change and Acute Conflict. Cited in Watagawa, “Unconventional Security Threats.”

———. 2000. *World's Water 2000–2001: The Biennial Report on Freshwater Resources*. Washington, DC: Island Press.

———. 2003. “A Soft Path: Conservation, Efficiency and Easing Conflicts over Water.” In *Whose Water Is It? The Unquenchable Thirst of a Water Hungry World*, edited by B. McDonald and D. Jehl. Washington, DC: National Geographic Society.

———. 2006. “Water and Terrorism.” In *The World's Water*, edited by P. H. Gleick et al. Washington, DC: Island Press.

Godement, F. 2009. “China Steps Up to World Stage, Cautiously.” April 24. http://yaleglobal.yale.edu/article.print?id=12295.

Goldenberg, S. 2009. "China Takes First Step towards Climate Deal by Committing to Carbon Target." *Guardian*, September 22.

Goldsmith, E., and N. Hildyard, eds. 1984. *The Social and Environmental Impacts of Large Dams*. Vol. 1. Cornwall, UK: Wadebridge Ecological Centre.

Goldstone, J. A. 2010. "The New Population Bomb: The Four Megatrends That Will Change the World." *Foreign Affairs*, January–February, pp. 31–43.

Gopinath, M., S. Dasgupta, and N. Surendran. 2003. *Conflict Resolution: Trends and Prospects*. New Delhi: WISCOMP.

Gordon, P. H. 2007. "Pakistan at Cross Roads—Pt. I." http://yaleglobal.yale.edu/display.article?id=9521.

Gore, A. 2006. *Inconvenient Truth*. Emmaus, PA: Rodale.

Gounder, R., and V. Xayavong. 2001. *Globalization and the Island Economies of South Pacific*. Helsinki: United Nations University.

Government of India. 2004. *A Reference Manual*. New Delhi: Publication Division, Ministry of Information and Broadcasting.

Grammaticas, D. 2009. "China's Climate Change Paradox." Accessed December 21. http://www.abc.net.au/lateline/content/2009/s2776467.htm.

Gready, P. 2004. *Fighting for Human Rights*. London: Routledge.

Greenfield, G. 2000. "The World Bank's 'Effective State' in East Asia." In Tang, *Social Development in Asia*.

Greenwald, C., and J. Kahn. 2009. *Globalization: The Irrational Fear That Someone in China Will Take Your Job*. Hoboken, NJ: Wiley.

Grynberg, R. 1998. *Rules of Origin: Issues in Pacific Island Development*. Canberra: Australian National University.

Guest, A. 1995. "Conflict and Cooperation in a Context of Change: A Case Study of the Senegal River Basin." In *Boundaries in Question: New Directions in International Relations*, edited by J. Macmillan and A. Linklater. London: Pinter.

Gupta, H. K. 1992. *Reservoir-Induced Earthquakes*. Amsterdam: Elsevier.

Gupta, J. 2010. "Sunderbans Will Drown in 60 Yrs: WWF." *Times of India*, February 22.

Haldar, C. 2004. "A Himalayan Effort: A Literary Review Article—India-Nepal Relations: The Challenge Ahead." *Statesman Weekly*, September 4.

Hansen, J. 2006a. "The Planet in Peril, Parts I and II." *Yale Global*, October 19 and 24.

———. 2006b. "The Threat to the Planet." *New York Review of Books*, July 13.

———. 2009a. "Can We Reverse Global Climate Change? Part I." May 14. http://yaleglobal.yale.edu/display.article?id=12371.

———. 2009b. "Climate Scientist Discusses Copenhagen Summit." December 7. http://www.abc.net.au/lateline/content/2008/s2764523.htm.

Haqqani, H. 2007. "Pakistan: Nuclear Power with Feet of Clay." May 22. http://yaleglobal.yale.edu.display.article?id=9206.

———. 2008. "Democracy in Pakistan Might Bring Tension with Washington." February 22. http://yaleglobal.yale.edu/display.article?id=10407.

Harrabin, R. 2007. "How Climate Change Hits India's Poor." http://news.bbc.co.uk/2/hi/south_asia/6319921.stm.

Hastings, A., and D. B. Wysham. 2010. "Regional Shifts in Ecological Systems Can Occur with No Warning." *Ecology Letters*. doi: 10.1111/j.1461-0248.2010.01439.x.

Haugen, Gary, and Victor Boutros. 2010. "And Justice for All: Enforcing Human Rights for the World's Poor." *Foreign Affairs*, May/June.

Haviland, C. 2007. "Nepal—Rocky Path to Democracy?" http://news.bbc.co.uk/go/pr/fr/-/2/hi/south-asia/6266575.stm.

Heap, P., B. Carin, and G. Smith. 2009. "Beyond Copenhagen: How to Cool the Planet." *Yale Global*, October 21.

Hendrie, Doug. 2010. "Why We're Doomed on Climate Change." Accessed January 7. http://www.abc.net.au/unleased/42180.html.

Hewett, Andrew. 2010. "Waking Up from the Copenhagen Hangover." Accessed January 7. http://www.abc.net.au/unleashed/42166.html.

Hiro, D. 2010. "New World Order Without a Hegemon: Compete and Cooperate." *Yale Global*, February 24.

Hitchens, C. 2007. *God Is Not Great: How Religion Poisons Everything*. Sydney, Australia: Allen & Unwin.

Hoffmann, S. A. 1998. "The International Politics of Southern Asia." In *Zones of Amity, Zones of Enmity: The Prospects for Economic and Military Security in Asia*, edited by J. Sperling, Y. Malik, and D. Louscher. Boston: Brill, Leiden.

Houghton, Sir John. 2007. *Global Warming: The Complete Briefing*. 3rd ed. Cambridge, UK: Cambridge University Press.

Howden, D. 2007. "Deforestation: The Hidden Cause of Global Warming." *The Independent*, May 15.

Huang, Y., and D. L. Yang. 2004. "Population Control and State Coercion in China." In *Holding China Together: Diversity and National Integration in the post-Deng Era*, edited by B. J. Naughton and D. L. Yang. Cambridge, UK: Cambridge University Press.

Huggler, J. 2007. "Retreating Himalayan Icefields Threatening Drought in Bangladesh." *Independent*, March 29.

Hundley, N., Jr. 1966. *Dividing the Waters: A Century of Controversy between the United States and Mexico*. Berkeley: University of California Press.

Hunter, J. M., L. Rey, K. Y. Chu, E. O. Adekolu-John, and K. E. Mott. 1993. *Parasitic Diseases in Water Resource Development: The Need for Intersectoral Negotiation*. Geneva, Switzerland: World Health Organization.

Hurrell, A. 2006. "Hegemony, Liberalism, and Global Order: What Space for Would-Be Powers." *International Affairs* 82 (1).

Hussain, W. 2003. "Bangladeshi Migrants in India: Towards a Practical Solution—A View from the North-Eastern Frontier." In *Missing Boundaries: Refugees, Migrants, Stateless and Internally Displaced Persons in South Asia*, edited by P. R. Chari, M. Joseph, and S. Chandran. New Delhi: Manohar.

Hussein, M. 2004. "Nationalities, Ethnic Processes and Violence in India's North East." In *Peace Studies: An Introduction to the Concept, Scope and Themes*, edited by R. Samaddar. New Dehli: Sage.

Iggers, G. G., and Q. E. Wang. 2008. *A Global History of Historiography*. London: Pearson-Longman.

Ingram, H. 1990. *Water Politics: Continuity and Change*. Albuquerque: University of New Mexico Press.

Ingram, H., D. Fieldman, and J. M. Whiteley. 2008. "Water and Equity in a Changing Climate." In *Water, Place and Equity*, edited by J. M. Whiteley, H. Ingram, and R. Perry. Cambridge, MA: MIT Press.

Ingram, H., J. M. Whiteley, and R. Perry. 2008. "The Importance of Equity and the Limits of Efficiency in Water Resources." In *Water, Place and Equity*, edited by J. M. Whiteley, H. Ingram, and R. Perry. Cambridge, MA: MIT Press.

Intergovernmental Panel on Climate Change. 2007. "Summary for Policymakers." In *Climate Change 2007: The Physical Science Basis. Contribution of Working Group 1 to the Fourth Assessment Report of the Intergovernmental Panel on Climate Change*, edited by S. Solomon, D. Qin, M. Manning, et al. Cambridge, UK: Cambridge University Press.

International Rice Research Institute. 2004. July 1. http://story.news.yahoo.com/news?tmpl.

Irani, C. R. 2005. "CAVEAT—Utter Lack of Conscience!" *Statesman*, May 1.

Iyer, R. R. 2000. *Water: Charting a Course for the Future*. New Delhi: Centre for Policy Research.

———. 2002. "Water Conflicts: A Note." In *Water Resources and Sustainable Development*, edited by K. Prasad. New Delhi: Shipra.

———. 2003. *Water: Perspectives, Issues, Concerns*. New Delhi: Sage.

James, H. 2009. "The Making of a Mess." *Foreign Affairs* 88 (1).

Jensen, M. E., W. R. Rangeley, and P. Dieleman. 1986. "Irrigation Trends in World Agriculture." In *Irrigation of Agricultural Crops*, edited by B. A. Stewart and D. R. Nielson. Madison, WI: American Society of Agronomy.

Jhaveri, N. 1988. "Three Gorges Debacle." *Ecologist* 18 (2–3).

Jiang, W. 2009. "China Tries to Wriggle out of the US Dollar Trap." April 29. http://yaleglobal.yale.edu/display.article?id=12311.

Kahlown, M. A., A. D. Khan, and M. Azam. 2006. "The World's Largest Contiguous Irrigation System: Developments, Successes and Challenges of the Indus Irrigation System in Pakistan." In Tvedt and Jakobsson, *History of Water*.

Kahn, J. 2007. "In China, Talk of Democracy Is Simply That." *New York Times*, April 20.

Kamdar, M. 2007. "Climate-Change Challenge for the Poor—Part II." Accessed January 12, 2008. http://yaleglobal.yale.edu/article.print?id=9734.

Kaplan, Robert D. 2010. "The Geography Power." *Foreign Affairs*, May/June.

Karhu, K., H. Fritze, H. Hamalainen, et al. 2010. "Temperature Sensitivity of Soil Carbon Fractions in Boreal Forest." *Soil Ecology* 91: 370–76.

Kathuria, V., and H. Gundimeda. 2002. "Industrial Pollution Control: Need for Flexibility." India Development Report 141. Oxford, UK: Oxford University Press.

Katz, D. 2006. "Going with the Flow: Preserving and Restoring Instream Water Allocations." In *The World's Water*, edited by P. H. Gleick et al. Washington, DC: Island Press.

Kelleher, E. 2008. "An Appetite for Food Stocks." *Financial Times* (UK), January 26.

Keller, W. W., and T. G. Rawski. 2007a. "Asia's Shifting Strategic and Economic Landscape." In *China's Rise and the Balance of Influence in Asia*, edited by W. W. Keller and T. G. Rawski. Pittsburgh: University of Pittsburgh Press.

———. 2007b. "China's Peaceful Rise: Road Map or Fantasy." In *China's Rise and the Balance of Influence in Asia*, edited by W. W. Keller and T. G. Rawski. Pittsburgh: University of Pittsburgh Press.

Kelly, M. 2003. "Science and Technology: The Essential Guide—Climate Change." Accessed November 24. http://www.bbc.co.uk/worldservice/sci_tech/features/essentialguide/theme_env.shtml.

Khadka, N. S. 1994. *Politics and Development in Nepal*. Jaipur, India: Nirala.

Khan, I. 2008. "An Interview with the Australian Broadcasting Commission 'Breakfast' Program." January 25. http://www.abc.net.au/rn/breakfast.

Khanal, K. 2007. "The New Battlefront (Nepal's Ethnic Politics)." *Economist*, March 10.

King, L. C. 1983. *Wandering Continents and Spreading Sea Floors on an Expanding Earth*. Chisester, UK: Wiley.

Klare, M. T. 2001. *Resource Wars: The Changing Landscape of Global Conflict*. New York: Metropolitan Books.

———. 2010. "China's Global Shopping Spree: Is the World's Future Resource Map Tilting East?" http://www/grist.org/article/2020-o4-05-chinas-global-shopping-spree-is-the-worlds-future-resource-map-t.

Kugelman, M., and S. L. Levenstein. 2010. "Sacrficing the Environment for Food Security." *World Political Review*, January 20. http://www.worldpoliticsreview.com/article.aspx?id=4969.

LaFraniere, S., and J. Grobler. 2009. "China Spreads Aid in Africa, with a Catch." *New York Times*, September 21.

Lancaster, J. 2003. "Peace Hopes Dim Again in Nepal." *Washington Post*, November 7.

Landay, J. S. 2008. "Insurgencies Spread in Afghanistan and Pakistan." *McClatchy Newspapers*, February 3.

LaPorte, R., Jr. 1972. "Pakistan in 1972: The Disintegration of a Nation." *Asia Survey* 12 (2).

Leake, J. 2010. "Natural Disasters Not Linked to Global Warming." *Sunday Times* (London), January 25.

Lean, G. 1999. "The Hidden Tentacles of a Super Power." *The Independent*. Reprinted in the *Canberra Times*, August 7.

Lehmann, Jean-Pierre. 2010. "Fallout from Copenhagen: Has the EU Lost Its Global Relevance?" Accessed January 11. http://yaleglobal.yale.edu/content/fallout-copenhagen-has-eu-lost-its-global-relevance.

Lewis, J. P. 1995. *India's Political Economy: Governance and Reform*. Oxford, UK: Oxford University Press.

Libiszewski, S. 1994. "Water, Water, Everywhere . . . " *The Ecologist* 24 (5).

———. 1999. "International Conflicts over Freshwater Resources." In *Ecology, Politics and Violent Conflict*, edited by M. Suliman. London: Zed Books.

Linter, B. 2010. "Northeast India: Boiling Pot of International Rivalry—Parts I and II." Accessed February 20. http://yaleglobal.yale.edu/content/northeast-india-boiling-pot-international-rivalry-part-i.

Littleton, M. 2009. "The TRIPs Agreement and Transfer of Climate-Change Related Technologies to Developing Countries." *Natural Resource Forum* 33: 233–34.

Lomborg, B. 2010. "Spare Us the Scares." *Times of India*, February 27.

Lorenzoni, I., A. Jordon, M. Hulme, et al. 2000. "A Co-evolutionary Approach to Climate Change Impact Assessment: Part I. Integrating Socio-economic and Climate-Change Scenarios." *Global Environmental Change* 10: 57–68.

Louka, E. 2008. *Water Law and Policy: Governance without Frontiers*. Oxford, UK: Oxford University Press.

Lowi, M. R. 1999. "Transboundary Resolutions." In *Contested Grounds: Security and Conflict in the New Environmental Politics*, edited by D. H. Deudney and R. E. Mathew. Albany: State University of New York Press.

Lunt, D. J., A. M. Haywood, G. A. Schmidt, U. Salzmann, P. J. Valdes, and H. J. Dowsett. 2009. "Earth System Sensitivity Inferred from Pliocene Modelling and Data." *Nature Geoscience*, December 6. doi: 10.1038/ngeo706.

L'vovitch, M. I. 1979. "World Water Resources and Their Future." *Geojournal* 3 (5). Translated by L. Nace from Russian.

MacDonald, S. B. 2007. "China and India: Same Globalization Road, Different Destinies." http://yaleglobal.yale.edu/article.print?id=9871.

Macey, Jennifer. 2010. "Cancun Approves Carbon Storage as Offset Projects in Poor Countries." Accessed January 8, 2011. http://www.abc.net.au/pm/content/2010/s3095269.htm.

Mahmoud, F. B. 1999. "The Gender Impact of War, Environmental Disruption and Displacement." In *Ecology, Politics and Violent Conflict*, edited by M. Suliman. London: Zed Books.

Majumder, S. 2008. "India Worries over Nepal's Future." April 14. http://news.bbc.co.uk/I/hi/world/south_asia/7347227.stm.

Malley, William. 2007. "Is Pakistan on the Edge of an Abyss?" Paper presented at the Australian Institute of International Affairs, Canberra, November.

Margolis, E. S. 2006. "Beijing's African Summit: Why China Is Wooing the Dark Continent." *Khalej Times*, November 17.

Marull, Y. 2010. "Brazil Rushes Ahead with Giant Dam." *The Canberra Times*, April 22.

Mastanduno, Michael. 2003. "Economic Engagement Strategies: Theory and Practice." In *Economic Interdependence and Conflict: New Perspectives on an Enduring Debate*, edited by Edward D. Mansfield and Brian M. Pollins. Ann Arbor: University of Michigan Press.

Mathur, S. 2010. "India Keeps Copenhagen Pledge on Emission Cuts." *Times of India*, January 31.

Mazzetti, M., and E. Schmitt. 2009. "Afghan Strikes by Taliban Get Pakistan Help, US Aides Say." *New York Times*, March 23.

McAllister, D. E., A. L. Hamilton, and B. Harvey. 1997. "Global Freshwater Biodiversity: Striving for the Integrity of Freshwater Ecosystems." *Sea Wind—Bulletin of Ocean Voice International* 11 (3). Cited in Fishbone, *The Struggle for Water*, 18.

McCully, P. 1996. *Silenced Rivers: The Ecology and Politics of Large Dams*. London: Zed Books.

McGivering, J. 2009. "The Talebanisation of Pakistan." Accessed May 4, 2009. http://newsvote.bbc.co.uk/mpapps/pagetools/print/news.bbc.co.uk/I/hi/programmes/from_.

McMahon, S. M., G. G. Parker, and D. R. Miller. 2010. "Evidence for a Recent Increase in Forest growth." *Proceedings of the Smithsonian Environmental Research Centre* 107 (8): 3611–15.

McMichael, A., et al. 2001. "Human Health." In *Third Assessment Report*, by the Working Group II. Geneva, Switzerland: Intergovernmental Panel on Climate Change.

McMullen, A. 2008. "Food Crisis Looms, BMO Analyst Warns; China, India Demand; 'It's Going to Hit This Year Hard' Donald Coxe Says." *Financial Post*, January 4.

Meehl, G. 2009. "Report: Some Climate Damage Already Irreversible." As interviewed by R. E. Schmid. Accessed January 27, 2009. http://news.yahoo.com/s/ap/20090126/ap_on_sc/sci_greenhouse_irreversible_damage.

Melik, J. 2009. "Food Crisis Hits Developing World Farm." Accessed April 17, 2009. http://newsvote.bbc.co.uk/mpapps/pagetools/print/news.bbc.co.uk/2/hi/business/7905312.

Mendelsohn, R., A. Dinar, and A. Sanghi. 2001. "The Effect of Development on the Climate Sensitivity of Agriculture." *Environment and Development Economics* 6 (1).

Mendelsohn, R., A. Dinar, and L. Williams. 2006. "The Distributional Impact of Climate Change on Rich and Poor Countries." *Environment and Development Economics* 2.

Milanovic, B. 2008. "Developing Countries Worse Off Than Once Thought." *Yale Global*.

Miliband, E. 2009. "The Road from Copenhagen." *Guardian*, December 20.

Milliman, J. D., J. M. Broadus, and F. Gable. 1989. "Environment and Economic Implication of Rising Sea Level and Subsiding Deltas: The Nile and Bengal Examples." *AMBIO* 18 (6).

Mishra, P. 2007. "Exit Wounds: The Legacy of Indian Partition." *New Yorker*, August 13.

Mishra, R. 2007. "Nepal's Attention Turns to King." http://news.bbc.co.uk/go/pr/fr/2/hi/south_asia/6515533.stm.

Misra, S. D. 1970. *Rivers of India*. New Delhi: National Book Trust.

Mitra, A. K. 1996. "Irrigation Sector Reform." *Economic and Political Weekly* 31 (13).

Mohan, R. V. R. 2003. "Rural Water Supply in India: Trends in Institutionalizing People's Participation." *Water International* 28 (4).

Mollinga, P. P., R. Doraiswamy, and K. Engbersen. 2004. "Capture and Transformation: Participatory Irrigation Management in Andhra Pradesh." In *Politics of*

Irrigation Reform: Contested Policy Formulation and Implementation in Asia, Africa and Latin America, edited by P. P. Mollinga and A. Bolding. Aldershot, UK: Ashgate.

Moran, T. 1993. "An Economic Agenda for Neorealists." *International Security* 18: 211–15.

Morse, B., and T. Berger. 1992. *Sardar Sarovar: The Report of the Independent Commission*. Ottawa, ON: Resource Future International.

Motaal, D. A. 2009. "Awaiting Climate Accord, Governments Toy with Dubious Measures." *Yale Global*, November 6.

Moy, A. D., W. R. Howard, S. G. Bray, and T. W. Trull. 2009. "Reduced Calcification in Modern Southern Ocean Planktonic Foraminifera." *Nature Geoscience* 2: 276–80.

Mukherjee, S. 2001. "Water Resource Development and the Environment in Bhutan." In *Sustainable Development of the Ganges, Brahmaputra-Meghna Basins*, edited by A. K. Biswas and J. I. Uitto. Tokyo: United Nations University Press.

Müller-Kraenner, S., and M. Kremer. 2010. "Europe's Green Diplomacy: Global Climate Governance Emerges as Test Case for EU." *Yale Global*, March 11.

Munir-uz-Zaman, M. 1987. "Issues Relating to the Uses of International Water Resources and World Bank Practices." In *Water Resources Policy for Asia*, edited by M. Ali, G. E. Radosevich, and A. A. Khan. Rotterdam, Netherlands: A. A. Balkema.

Musil, R. K. 2009. *Hope for a Heated Planet*. New Brunswick, NJ: Rutgers University Press.

Mustafa, S. 2007. "Bangladesh at Crossroads." British Broadcasting Commission. April 5. http://news.bbc.co.uk/2/hi/south_asia/6530781.stm.

Myers, N. 1996. *Ultimate Security: The Environmental Basis of Political Stability*. Washington, DC: Island Press.

Nair, A. R., A. S. Pendharkar, S. V. Navada, and S. M. Rao. 1978. "Groundwater Recharge Studies in Maharastra." *Isotope Technology* 2 (International Atomic Energy Agency, Vienna).

Naqvi, N. 2008. "Why India? Why Now?" *Far Eastern Economic Review*, December.

Nasrawi, S. 2010. "Egypt Shuns Nile Sharing Deal with African Nations." *Arab News*, April 14. http://arabnews.com/middleeast/article43561.ece.

Nath, D. C. 2004. *Intelligence Imperatives for India*. New Delhi: India First Foundation.

Naughton, B. J. 2004. "Western Development Program." In *Holding China Together: Diversity and National Integration in the Post-Deng Era*, edited by B. J. Naughton and D. L. Yang. Cambridge, UK: Cambridge University Press.

Nayar, K. 1975. *Distant Neighours: A Tale of the Subcontinent*. Delhi: Vikash.

Newberg, P. R. 2006. "Will Guns Fall in South Asia?" December 5. http://yaleglobal.yale.edu/display.article.article?id=8502.

———. 2007. "A US Alliance with One Man." *Yale Global*, July 31.

———. 2008. "Will Ballots Win over Bullets in Pakistan?" *Yale Global*, February 15.

Nilekani, R. 2007. "Is Water the Next Oil?" May 31. http://yaleglobal.yale.edu/display.article?id=9243.

Nkrumah, G. 2004. "It Must Be Something in the Water: The Controversy over Nile: Water Distribution Is Rearing Its Ugly Head." *Al-Ahram Weekly*, March 3.

Norton, A. 2004. *Leo Strauss and the Politics of American Empire*. New Haven, CT: Yale University Press.

Nye, J. S., Jr., and J. D. Donahue, eds. 2000. *Governance in a Globalizing World*. Washington, DC: Brookings Institution Press.

Ohlsson, L. 1995. "Role of Water and the Origin of Conflict." In *Hydropolitics: Conflicts over Water as a Development Constraint*, edited by L. Ohlsson. Dhaka, Bangladesh: Dhaka University Press.

Olson, E. 2004. "Unnatural Weather, Natural Disasters: A New UN Focus." *New York Times*, May 18.

Oppenheimer, Michael. 2010. "IPCC Scientists on the Defensive as Sceptics Step Up Assault." Australian Broadcasting Commission. February 4. http://www.abc.net.au/7.30/content/2010/s2810662.htm.

Orlik, J., and W. Yong. 2009. "China Tells Europe to Mind Its Manners." *Far Eastern Economic Review*, March.

Oster, S. 2006. "Beijing Hosts Big Africa Summit as West Watches." *Wall Street Journal*, October 25.

———. 2009. "World's Top Polluter Emerges as Great-Technology Leader." *Wall Street Journal*, December 15. Reprinted in *Yale Global*, http://yaleglobal.yale.edu/content/worlds-top-polluter-emerges-green-technology-leader.

Pachauri, R. K. 2008. "The Hefty Cost of Climate Change." *Statesman*, January 25.

Pacific Institute, with United Nations Environment Programme. 2010. *Clearing the Waters: A Focus on Water Quality Solutions*. Nairobi, Kenya: Pacific Institute.

Padovani, F. 2006. "The Chinese Way of Harnessing Rivers: The Yangtze River." In Tvedt and Jakobsson, *History of Water*.

Pai, S. 2007. "Janadesh 2007: The Land Question." *Economic and Political Weekly* 42: 45–46.

Palaniappan, M., E. Lee, and A. Samulon. 2006. "Environmental Justice and Water." In *The World's Water*, edited by P. H. Gleick et al. Washington, DC: Island Press.

Pant, H. V. 2010. "Saving Afghanistan: Elephant in the Room Is Indo-Pakistan Rivalry." Accessed February 8. http://yaleglobal.yale.edu/content/solving-afghanistan-elephant-room-indo-pakistan-rivalry.

Parameswaran, P. 2007. "Pakistan's Election Already Rigged: UN." *Canberra Times*, December 15.

Patz, J. A., D. Campbell-Lendrum, T. Holloway, and J. A. Foley. 2005. "Impact of Regional Climate on Human Health." *Nature*, November 17. doi:10.1038/nature04188.

Pearce, F. 1992. *The Dammed: Rivers, Dams, and the Coming World Water Crisis*. London: Bodley Head.

———. 2004. "Rice Yields Plunging due to Balmy Nights." *New Scientists*. Accessed January 1, 2008. http://www.newscientist.com/article.ns?id=dn6082&print=true.

———. 2007. "The Big Barons." *Weekly New Scientist*, December 1.

Pearson, C., R. Blinch, and S. Kovalyova. (2009). "The Fight over the Future of Food." *Mail & Guardian*, December 10.

Pel ez, C. M., and C. A. Pel ez. 2008. *Globalization and the State*. Vol. 1. New York: Palgrave Macmillan.

Pelezero, C., E. Calvo, and O. Hoeg-Guldberg. 2010. "Paleo-perspectives on Ocean Acidification." *Trends in Ecology and Evolution.* Summarized in "Carbon Emissions Lead to Dangerous Changes in Oceans," *Times of India*, April 3, 2010.

Peterson, E. R., and P. B. Davies. 2005. *World's Water on Countdown*. Washington, DC: Centre for Strategic and International Studies.

Peterson, S. 2007. "Tensions on Iraq Border Rile Turkey." *Christian Science Monitor*, June 9.

Phadnis, U. 1990. *Ethnicity and Nation Building in South Asia*. New Delhi: Sage.

Phelps, J., E. L. Webb, and A. Agarwal. 2010. "Does REDD+ Threaten to Recentralize Forest Governance?" *Science* 238 (5976): 312–13.

Phillips, M. 2008. "New Partners in Pakistan." *Wall Street Journal*, February 22.

Pietz, D. A. 2006. "Controlling the Waters in Twentieth-Century China: The Nationalist State and the Huai River." In Tvedt and Jakobsson, *History of Water*.

Plaut, S. 2000. "Water Policy in Israel." Policy Studies 47. Jerusalem: Institute for Advance Strategic and Political Studies.

Polsky, C., and D. W. Cash. 2005. "Drought, Climate Change, and Vulnerability: The Role of Science and Technology in a Multi-scale, Multi-stressor World." In *Drought and Water Crises*, edited by D. A. Wilhite. Boca Raton, FL: Taylor & Francis.

Pomeranz, K. 2009. "The Great Himalayan Watershed: Water Shortages, Mega Projects and Environmental Politics in China, India and Southeast Asia." *Asia Pacific Journal*, July 27.

Postel, S. 1992. *Last Oasis (Facing Water Scarcity)*. London: Norton.

Postigo, A. 2008. "Vulnerability and Adaptation to the Health Impacts of Climate Change." *Development: Climate Justice and Development* 51 (3).

Pronk, J. 2008. "Climate, Scarcities and Development." *Development: Climate Justice and Development* 51 (3).

Prüss, A., D. Kay, L. Fewtrell, and J. Bartram. 2002. "Estimating the Burden of Disease from Water, Sanitation, and Hygiene at a Global Level." *Environmental Health Perspectives* 110: 537–42.

Puri, R. 2009. "Lesson from Lahore." *Statesman*, March 4.

Qureshi, S. 1981. "Military in the Policy of Islam: Religion as a Basis for Civil Military Reaction." *International Political Science Review* 2 (3).

Rachman, G. 2008. "The Battle for Food, Oil and Water." *Financial Times* (Asian edition), January 29.

Račius, E. 2007. *Muslim Missionary Activities Between Religion and Politics: The Multiple Nature of Islamic Da'wa*. Vilnius, Lithuania: Institute of International Relations and Political Science, Vilnius University.

Rahman, A. A. 1999. "Climate Change and Violent Conflict." In *Ecology, Politics and Violent Conflict*, edited by M. Suliman. London: Zed Books.

Rahman, W. 2006. "Is Bangladesh Winning against Extremism?" British Broadcasting Commission. March 6. http://news.bbc.co.uk/2/hi/south_asia/4779544.stm.

———. 2007. "Is Bangladesh Heading towards Disaster?" British Broadcasting Commission. January 8. http://news.bbc.co.uk/2/hi/south_asia/6241263.stm.

Ramakrishnan, T. 2004. "Sharing Water Resources." *The Hindu*. http://www.hinduonnet.com/2004/10/08/stories/2004100803561 00.

Rashid, Ahmed. 2000. *Taliban: Islam, Oil and the New Great Game in South Asia*. London: I. B. Tauris.

———. 2007a. "Emergency for Pakistan—and the World." http://wwwyaleglobal.yale.edu/article.print?id=9965.

———. 2007b. "Pakistan's Anniversary Wish—Justice." http://yaleglobal.yale.edu/display.article?id=9558.

———. 2008. "After *Bhutto*, Pakistan on Edge." January 1. http://yaleglobal.yale.edu/content/after-bhutto-pakistan-edge.

Rasmus, D. W., and R. Salkowitz. 2009. *Listening to the Future*. Holbonken, NJ: Wiley.

Ravikumar, N. 2008. "The Changing Chennaites." *Statesman*, January 25.

Ray, B. 1998. "Farakka Treaty: Can It Lead to a Better Management of Water Resources in Bangladesh and India?" *Statesman*, January 16 and 17.

———. 2003. *South Pacific Least Developing Countries: Towards Positive Independence*. Kolkata, India: Progressive.

———. 2006. "Corruption: Perceptions, Reality and Policy Implications." In *Corruption, Governance and Democracy in South Asia*, edited by R. J. May and B. Ray. Kolkata, India: Towards Freedom.

———. 2008a. "Global Conventions and Regulations on International Rivers: Implications for South Asia." In *Water First: Issues and Challenges for Nations and Communities in South Asia*, edited by K. Lahiri-Dutt and R. J. Wasson. New Delhi: Sage.

———. 2008b. *Water: The Looming Crisis in India*. Lanham, MD: Lexington Books.

Read, R. 2001. *Growth, Economic Development and Structural Transition in Small Vulnerable States*. Helsinki: United Nations University Press.

Rees, M. 2010. "Royal Society Turns 350." An interview with ABC's science reporter R. Williamson. April 24. http://www.abc.net.au/rn/scienceshow/stories/2010/2881446.htm.

Revenga, C., J. Brunner, N. Henninger, K. Kassem, and R. Payne. 2000. *Freshwater Systems*. Washington, DC: World Resource Institute.

Revkin, A. C. 2007. "Poor Nations to Bear Brunt as World Warms." *New York Times*, April 2.

———. 2009. "*NY Times* environment reporter joins *Lateline*." ABC. Accessed December 21. http://www.abc.net.au/lateline/content/2009/s2767053.htm.

Rhein, E. 2010. "How to Pass from Copenhagen to Mexico." Accessed January 23. http://www.euuractiv.com/en/climate-environment/pass-copenhagen-mexico/article-189105.

Rhode, D. 2007. "A Detour from a Battle against Terror." *New York Times*, November 6.

Rich, B. 1994. *Mortgaging the Earth: The World Bank, Environmental Impoverishment and the Crisis of Development*. Boston: Beacon Press.

Richardson, M. 2008 "Dominating the Headwater." *Japan Times*, March 31.

Ridgwell, A., and D. N. Schmidt. 2010. "Past Constraints on the Vulnerability of Marine Calcifiers to Massive Carbon Dioxide Release." *Nature Geoscience* 3: 196–200. doi: 10.1038/ngeo755.

Rizvi, Hasan-Askari. 1998. "Pakistan: Civil-Military Relations in a Praetorian State." In *The Military and Democracy in Asia and the Pacific*, edited by R. J. May and V. V. Selochan. Bathurst, New South Wales: Crawford House.

Robert, J. 1994. *Water Is a Commons*. Mexico City, Mexico: Habitat International Coalition.

Robinson, S. 2007. "Pakistan: The Making of a Crisis." *Time*. November 9. http://www.time.com/time/world/article/0,8599,1682534,00.html.

Rogelj, J., J. Nabel, C. Chen, et al. 2010. "Copenhagen Accord Pledges Are Paltry." *Nature* 464: 1126–28. doi: 10.1038/4641126a.

Rogers, A. D. 2010. "Adieu, Atlantic Blue Fin Tuna." *Yale Global*, March 17.

Rosegrant, M. W., Ximing Cai, and S. A. Cline. 2002. *World Water and Food to 2025*. Washington, DC: International Food Policy Research Institute.

Rosenthal, E. 2008. "World Food Stocks Dwindling Rapidly." *International Herald Tribune*, January 10.

Rothchild, D. 2002. "The Effects of State Crisis on African Interstate Relations (and Comparison with Post-Soviet Eurasia)." In *Beyond State Crisis? Post-Colonial Africa and Post-Soviet Eurasia in Comparative Perspective*, edited by M. R. Beissinger and C. Young. Washington, DC: Woodrow Wilson Center Press.

Rothfeder, J. 2001. *Every Drop for Sale*. New York: Tarcher/Putnam.

Roy, S. 2004. "Rise to Meet the Water Challenge." December 11. http://www.google.comau/search?q.cache:85bqRXpXMJ:inrediff.com/news/2004.

———. 2007. "On Indian Nationhood." *Statesman Weekly*, June 2.

Runge, C., and B. Senauer. 2007. "How Biofuels Could Starve the Poor." *Foreign Affairs* 86 (3).

Sabur, A. K. M. A. 1995. "South Asian Security in Post–Cold War Era: Issues and Outlook." In *South Asia's Security: Primacy of Internal Dimensions*, edited by Iftekharuzzaman. Dhaka, Bangladesh: Academic.

Sachs, W. 2008. "Climate Change and Human Rights." *Development: Climate, Justice and Development* 51 (3).

Salman, M. A. S., and L. B. de Chazournes. 1998. *International Watercourses: Enhancing Cooperation and Managing Conflicts*. Proceedings of a World Bank Seminar. Washington, DC: World Bank.

Salman, M. A. S., and K. Uprety. 2002. *Conflicts and Cooperation on South Asia's International Rivers*. The Hague, Netherlands: Kluwer Law International.

Santhanam, R. 2007. "A Question of Survival." *Statesman*, November 5.

Saunders, P. J., and V. Turekian. 2007. "Why Climate Change Can't Be Stopped." *Foreign Affairs*, October 18. http://yaleglobal.yale.edu/display.article?id=9830.

Sautman, B., and Y. Hairong. 2010. "Demonizing China: Pundits Get Its Role in Africa Wrong." http://yaleglobal.yale.edu/content/demonizing-china-pundits-get-its-role-in-africa-wrong.

Schell, O. 2009. "China Reluctant to Lead." March 11. http://yaleglobal.yale.edu/display.article?id=12074.

Schindler, D. 2003. "Mountain Water: Lifeblood of the Prairies." In *Whose Water Is It? The Unquenchable Thirst of a Water Hungry World*, edited by B. McDonald and D. Jehl. Washington, DC: National Geographic Society.

Schwartz, P., and D. Randall. 2003. "Abrupt Climate Change Scenario and Its Implications for US National Security." http://www.ems.org/climate/exec_pentagon_climatechange.pdf.

Science. 2010. "Climate Change and the Integrity of Science." 328: 689–90.

Science and Technology Commission, People's Republic of China. 1991. "Beijing-Tianjin Water Resources Study." Beijing. Cited in Postel, *Last Oasis*, 28–37.

Searchinger, T., R. Heimlich, R. A. Houghton et al. 2008. "Use of U.S. Croplands for Biofuels Increases Greenhouse Gases through Emissions from Land-Use Change." *Science* 319 (5867): 1238–1240.

Segal, A. 2007. "Chinese Economic Statecraft and the Political Economy of Asian Security." In *China's Rise and the Balance of Influence in Asia*, edited by W. W. Keller and T. G. Rawski. Pittsburgh: University of Pittsburgh Press.

Sen, A. 1981. *Poverty and Famine: An Essay on Entitlement and Deprivation*. Oxford, UK: Oxford University Press.

Sengupta, N. 1985. "Irrigation: Tradition vs. Modern." *Economic and Political Weekly* 20 (45–47).

Sen Sharma, S. 1997. "Large Dam Projects of India—An Evaluation." *Science and Culture* 63.

Serageldin, I. 1995. "If the Wars of the Twentieth Century Were Fought over Oil, the Wars of This Century Will Be Fought over Water." *New York Times*, August 10.

Shambaugh, D. 1996. "Containment or Engagement of China? Calculating Beijing's Responses." *International Security* 21: 180–209.

Shearman, D., and J. W. Smith. 2007. *The Climate Change, Challenge and the Failure of Democracy*. Westport, CT: Praeger.

Sherpa, D. M. 1994. *Living in the Middle: Sherpas of the Middle-Range Himalayas*. Prospect Heights, IL: Waveland Press.

Shih, Chih-yu. 1990. *The Spirit of Chinese Foreign Policy: A Psychological View*. Hampshire, UK: Houndmills.

Shiva, V. 2003. *Sujalam: Living Waters—The Impact of the River Linking Project*. New Delhi: Navdanya.

Simon, P. 1998. *Tapped Out*. New York: Welcome Rain.

Simons, C. 2007. "Good Oil, Bad Environmental Effect, Ugly Cost." *Sydney Morning Herald*, June 19. http://www.smh.com.au/news/environment/good-oil-bad-environmental-effect-ugly-cost/.

Singh, A. 2009. "When SC Prioritised Public Interest." Accessed December 31. http://www.thestatesman.net/page.print.php?usrsess=1&id=279721&clid=2.

Singh, J. 1998. "South Asian Security Issues in the 1990s: Continuity and Change." In *Security Issues in South Asia: Domestic and External Sources of Threat to Security*, edited by A. K. Banerjee. Kolkata, India: Minerva.

Sivaraksa, S. 1993. *When Loyalty Demands Dissent: Sulak Sivaraksa and the Charge of Lese Majeste in Siam 1991–1993*. Bangkok: Santi Pracha Dhamma Institute.

Smith, J., and D. Shearman. 2006. *Climate Change Litigation*. Adelaide, Australia: Presidian.

Sobhan, R. 2000. "The Future of SAARC: Towards Building a South Asian Community." In *Economic Liberalization and Institutional Reforms in South Asia: Recent Experiences and Future Prospects*, edited by D. K. Dutta. New Delhi: Atlantic.

Solomon, S., Gian-Kasper Plattner, R. Knutti, and P. Friedlingstein. 2009. "Irreversible Climate Change due to Carbon Dioxide Emissions." National Academy of Science of the USA. Accessed March 9. http://www.pnas.org/content/106/6/1704.full.

Someshawr, S. 2008. "Adaptation as 'Climate-Smart' Development." *Development: Climate Justice and Development* 51 (3).

Sorman, G. 2007. "The Truth about China." *Wall Street Journal*, April 20.

Sperling, J. 1998. "Introduction: A Problem of Cooperation in the Asia-Pacific." *International Studies in Sociology and Social Anthropology* 70: 2–19.

Srinivasan, R. 2010. "No Financial Crisis Impact? India's Poor Grew by 34 Mn." *Times of India*, February 10.

Stavins, R. 2009. "What Hath Copenhagen Wrought? A Preliminary Assessment of the Copenhagen Accord." Accessed December 30. http://groups.google.com/geoengineering/browse_thread/thread/a01cfaccb982b6ac.

Steiner, Achim. 2010. Foreword. In *Water Quality Challenges and Solutions the Focus of World Water Day 2010 Report Awareness and Will to Change are Key*. Oakland, CA: Pacific Institute. http://www.pacinst.org/reports/water_quality/index.htm.

Stephenson, J. 2009. "Testimony before the Subcommittee on Commerce, Trade and Consumer Protection, Committee on Energy and Commerce, House of Representatives: Options for Enhancing the Effectiveness of the Toxic Substances Control Act." US Government Accountability Office. Cited in Pacific Institute, *Clearing the Waters*.

Stern, N. H. 2006. "Executive Summary." In *Stern Review: Economics of Climate Change*. London: H. M. Treasury.

———. 2009. "Climate Change Economist Nicholas Stern Joins *7.30 Report*." Accessed June 2. http://www.abc.net.au/7.30/content/2009/s2582625.htm.

Strauss, S. D. 2002. *The Complete Idiot's Guide to World Conflicts*. Indianapolis, IN: Alpha.

Strong, M., and G. Goransson. 1991. "Foreword." In *Water: The International Crisis*, edited by R. Clarke. London: Earthscan.

Suri, K. 2008. "Ladakh Feels Warmth of Green Revolution." *Statesman*, January 28.

Tanaka, A. 1995. "China: Dominant, Chaotic or Interdependent." In *Future Sources of Global Conflict*, edited by A. Tanaka and S. Sato. London: Royal Institute of International Affairs.

Tang, K. L. 2000a. "Moving Toward Xiaokang in China." In *Social Development in Asia*, edited by K. L. Tang. Dordrecht, Netherlands: Kluwer Academic.

——, ed. 2000b. *Social Development in Asia*. Dordrecht, Netherlands: Kluwer Academic.

Tanham, G. K. 1992. *Indian Strategic Thought: An Interpretive Essay*. Santa Monica, CA: RAND.

Tanzi, A., and M. Arcari. 2001. *The UN Convention on the Law of International Watercourses*. The Hague, Netherlands: Kluwer Law International.

Tellis, A. J. 2010. "Pakistan and the Afghanistan End Game—Part II." *Yale Global*, March 15.

Timmer, C. Peter. 2011. "Failed Policies Lead to Food Shortages." *Wall Street Journal*, February 22.

Tisdale, C. 2008. "Economic and Business Relations between China and Australia: Insights into China's Global Economic Footprint." In *Engagement and Change*, edited by P. Basu, G. O'Neill, and A. Travaglione. Brisbane: Australian Academic Press.

Trenberth, K. E., and J. T. Fasullo. 2010. "Tracking Earth's Energy." *Science*, April 16.

Tvedt, Terje, and Eva Jakobsson, eds. 2006. *History of Water*. Vol. 1. New York: I. B. Tauris.

United Nations. 1986. *Ground Water in Continental Asia*. Natural Resources/Water Series 15. New York: United Nations.

———. 1995. *Guidebook to Water Resources, Use and Management in Asia and the Pacific*. Vol. 1. New York: United Nations.

———. 1995–2008. *United Nations Human Development Report*. New York: United Nations.

United Nations Conference on Trade and Development. 1989. *Africa's Commodity Problems: Towards a Solution.*, Geneva, Switzerland: Task Force on United Nations Program of Action for African Economic Recovery and Development.

United Nations Development Programme. 2007/2008. "Fighting Climate Change: Human Solidarity in a Divided World." http://hdr.undp.org/reports/global/hdr2007-2008/.

United Nations Economic and Social Council. 2004. E/CN. 17/2004/4. 12th session, April 24–30.

United Nations Environment Programme. 2007. "Millions at Risk of Hunger and Water Stress in Asia Unless Global Greenhouse Emissions Cut." April 10. http://global.factiva.com/ha/default.aspx.

United Nations World Water Development Report. 2003. *Water for People, Water for Life*. New York: UNESCO.

Upadhyaya, S. P., and B. N. Sapkota. 1987. "Water Resources Development in Nepal." In *Water Resources Policy for Asia*, edited by M. Ali, G. E. Radosevich, and A. A. Khan. Rotterdam, Netherlands: A. A. Balkema.

Van Halsema, G. E., and L. Vincent. 2006. "Of Flumes, Modules and Barrels: The Failure of Irrigation Institutions and Technologies to Achieve Equitable Water Control in the Indus Basin." In Tvedt and Jakobsson, *History of Water*.

van Kemenade, W. 2008. "China and Pakistan: New Friends Can't Compare." *Yale Global*, March 12.

Varshney, A. 2007. "India's Democratic Challenges." *Foreign Affairs* 86 (2).

Vickers, A. 2005. "Managing Demand: Water Conservation as a Drought Mitigating Tool." In *Drought and Water Crises*, edited by D. A. Wilhite. Boca Raton, FL: Taylor & Francis.

Vidal, J. 2007a. "Biofuelling the Food Crisis." *Mail and Guardian*, September 10.

———. 2007b. "Global Rush to Energy Crops Threatens to Bring Food Shortages and Increase Poverty, Says UN." *The Guardian*, May 25.

Vines, A. 2007. "China in Africa: A Mixed Blessing." *Current History—Journal of Contemporary World Affairs*, May.

von Tunzelmann, Alex. 2007. *The Secret History of the End of an Empire*. New York: Holt.

Vőrősmarty, C. J., P. Green, J. Salisbury, and R. B. Lammers. 2000. "Global Water Resources: Vulnerability from Climate Change and Population Growth." *Science* 289 (5477).

Wade, R. 1985. "Market for Public Office: Why the Indian State Is Not Better at Development." *World Development* 13 (4).

———. 1988. *Village Republics: Economic Conditions for Collective Action in South Asia*. Cambridge, UK: Cambridge University Press.

Waldrop, M. M. 1992. *Complexity: The Emerging Science at the Edge of Order and Chaos*. New York: Simon & Schuster.

Wallerstein, I. 2008. *The Demise of Neo-liberal Globalization*. New York: Fernand Braudel Centre, Bingham University.

Wallis, W. 2008. "Drawings Contours of a New World Order." *Financial Times* (UK), January 24.

Ward, C. 1997. *Reflected in Water: A Crisis of Social Responsibility*. London: Cassell.

Watagawa, R. 1995. "Unconventional Security Threats: An Economist's View." In *Future Sources of Global Conflict*, edited by T. Taylor and S. Sato. Vol. 4. London: Royal Institute of International Affairs.

Webster, B., and M. Franklin. 2010. "Be Truthful on Climate: Science Boss." *Australian*, January 28.

Weitz, R. 2009. "Global Insights: Chavez Trades Oil for Arms in Moscow." *World Politics Review*, September 15

Werker, E. 2008. "Power to the People." *Foreign Policy*, November/December.

Whitcombe, E. 1972. *Agrarian Conditions in Northern India*. Vol. 1 of *The United Province under British Rule, 1860–1900*. Berkeley: University of California Press.

White, C. 2004. "Three Gorges Dam." *Discovery*. Accessed November 12, 2007. http://www.ccds.charlotte.nc.us/History/China/02/Cwhite/Cwhite.htm.

Wiarda, H. J. 2003. *Civil Society: The American Model and Third World Development*. Cambridge, MA: Westview Press.

Wilhite, D. A., and R. S. Pulwarty. 2005. "Drought and Water Crises: Lessons Learned and the Road Ahead." In *Drought and Water Crises*, edited by D. A. Wilhite. Boca Raton, FL: Taylor & Francis.

Wines, M. 2009. "Worst Drought in Half Century Shrives the Wheat Belt of China." *New York Times*, February 24.

Wolf, A., A. Kramer, A. Carius, and G. Dabelco. 2009. "Peace in the Pipeline (a Viewpoint)." Accessed February 16. http://news.bbc.co.uk/2/hi/science/nature/7886646.stm.

Wolf, A., J. A. Natharius, J. J. Danielson, B. S. Ward, and J. K. Pender. 1999. "International River Basins of the World." *International Journal of Water Resource Development.*

World Bank. 2000, 2002, 2003, 2005, 2006. *World Development Indicators*. Washington, DC: World Bank.

———. 2000, 2002, 2003, 2005, 2006. *World Development Report*. Washington, DC: World Bank.

———. 2004. *Water Resources Sector Strategy*. Washington, DC: World Bank.

World Health Organization. 2008a. "Burden of Disease and Cost-Effectiveness Estimates." Accessed December 29. http://www.who.int/water_sanitation_health/diseases/burden/en/print.html.

———. 2008b. "WHO and the Global Food Security Crisis." Accessed December 12. http://www.who.int/food_crisis/global_food_crisis/en/print.html.

World Resources. 1992. *World Resources 1992–93: A Report in Collaboration with UNEP and UNDP*. New York: Oxford University Press.

———. 1998. *World Resources 1998–99: A Guide to the Global Environment, Environmental Change and Human Health*. New York: Oxford University Press.

World Water Council. 2007. "The Use of Water Today." In *The Struggle for Water*, edited by A. Fishbone. London: International Debate Education Association.

Worship, P. 2009. "2008 Confirmed Rise in Weather Disasters—Red Cross." Accessed June 18. http://uk.reuters.com/articlePrint?articleId=UKN16299845._CH_.2420.

Xenopoulos, M., D. Lodge, J. Alcamo, M. Mårker, K. Schulz, and D. van Vuuren. 2005. "Scenarios of Freshwater Fish Extinctions from Climate Change and Water Withdrawal." *Global Change Biology* 11: 1–8. Cited in Katz, "Going with the Flow."

Yahuda, M. 2002. "China and Regional Cooperation." In *China's Place in Global Geopolitics: International, Regional and Domestic Challenges*, edited by K. E. Brodsgaard and B. Heurlin. New York: RoutledgeCurzon.

Yardley, J. 2004a. "China's Premier Orders Halt to a Dam Project Threatening a Lost Eden." *New York Times*, September 9.

———. 2004b. "Farmers Being Moved Aside by China's Real Estate Boom." *New York Times*, December 8.

———. 2007. "Beneath Booming Cities, China's Future Is Drying Up." *New York Times*, October 2.

Yas-Meen, S. 2007. "Bhutto Murder Sparks Instability Concerns in Pakistan." Interview with the Australian Broadcasting Commission. *7.30 Report*, December 28.

Yin, J., M. E. Schlesinger, and R. J. Stouffer. 2009. "Model Projections of Rapid Sea-Level Rise on the North-east Coast of the United States." *Nature Geoscience*. doi:10.1038/ngeo462.

Yu, Jianrong. 2007. "Social Conflict Is the Taboo of Chinese Political Discourse." *China Security* 3 (2): 2–17.

Zedillo, E. 2007. *Forbes*, March 19. http://global.yale.edu/display.article?id=8930.

Zhang, Hai Lun, Li Dan Ke, and Fa Shi Zhang. 2005. "Drought and Water Management: Can China Meet Future Demand?" In *Drought and Water Crisis: Science, Technology and Management Issues*, edited by D. A. Wilhite. Boca Raton, FL: Taylor & Francis..

Zhao, Huasbeng. 2005. *New Situation in Central Asia and Shanghai Cooperation Organization in China and Asia's Security.* China in the Globalized World Series 1. Singapore: Marshall Cavendish Academic.

Additional References for Further Reading

Acharya, A. 2008. *China and India: Politics of Incremental Engagement*. New Delhi: Har-Anand.

Ahmed, M. 2008. "Generals Urge Musharraf Resignation." January 22. http://www.time.com/time/world/article/08599,1706102,00.htm.

———. 2009. "Samina Ahmed Discusses Refugee Exodus in Pakistan." May 11. http://www.abc.net.au/lateline/content/2008/s2567395.htm.

Alavi, H. 1983. "Class and State." In *Pakistan: The Root of Dictatorship—The Political Economy of a Praetorian State*, edited by H. N. Gardezi and J. Rashid. London: Zed Press.

Blackwill, R. D. 2009. "Pakistan, Taliban and Global Security—Part II." May 12. http://yaleglobal.yale.edu/display.article?id=12361.

Brown, L. R. 1980. *Food for Fuel: New Competition for the World's Cropland*. Worldwatch Paper 35. Washington, DC: Worldwatch Institute.

Chakraborty, R. 2004. "Sharing of Water among India and Its Neighours in the 21st Century: War or Peace?" *Water International* 29 (2).

Chanda, N. 2007. "Pakistan at the Crossroads." http://yaleglobal.yale.edu/display.article?id=9534.

Chellaney, B. 1994. "India." In *Nuclear Proliferation after the Cold War*, edited by M. Reiss and R. S. Litwak. Washington, DC: Woodrow Wilson Centre Press.

Chrislip, D. D., and C. E. Larson. 1994. *Collaborative Leadership*. San Francisco: Jossey-Bass.

Clare, M. T. 2001. *Resource Wars: The Changing Landscape of Global Conflict*. New York: Metropolitan Books.

Cloughley, B. 2006. *A History of the Pakistan Army: Wars and Insurrections*. 3rd ed. Oxford, UK: Oxford University Press.

Commonwealth Scientific and Industrial Research Organisation. 2008. "Human-Generated Aerosols Affect Our Weather." CSIRO Marine and Atmospheric Research. http://www.csiro.gov.au.

Connelly, P., and R. Perlman. 1975. *Politics of Society: Resource Conflicts in International Relations*. London: Royal Institute of International Affairs.

Disraeli, B. 1866. *Wit and Wisdom of Benjamin Disraeli: Collected from his Writings and Speeches*. New edition, 1883. London: Longmans, Green.

Economist. 2010. "Planet B: How the Underwhelming Copenhagen Accord Could Yet Turn into a Useful Document." Accessed January 23. http://www/economist.com/opinion/displaystory.cfm?story_id=15174523.

Feldman, D. L. 1991. *Water Resources Management: In Search of an Environmental Ethic*. Baltimore: Johns Hopkins University Press.

Ganne, E. 2009. "More Carrot and Less Stick Needed to Fight Global Warming." *Yale Global*, October 28.

Gleick, P. H. 1993. *Water in Crisis: A Guide to Fresh Water Resources*. Oxford, UK: Oxford University Press.

Gossman, P. 2002. "Kashmir and International Law: How War Crimes Fuel the Conflict." Accessed January 16, 2004. http://www.crimesofwar.org/onnews.

Harries, O. 2005. *Morality and Foreign Policy*. Occasional Paper 14. Sydney: Centre for Independent Studies.

Harshe, R. 2001. "Understanding Conflicts in South Asia." In *Intra and Inter-state Conflicts in South Asia*, edited by S. G. George. New Delhi: South Asian.

Homer-Dixon, T. F., and J. H. Boutwell. 1993. "Environmental Change and Violent Conflict." *Scientific American* 268 (2).

Houghton, Sir John. 2007. *Global Warming: The Complete Briefing*. Cambridge, UK: Cambridge University Press.

Kamal, A. 2006. "Living with Water: Bangladesh since Ancient Times." In *History of Water*, edited Terje Tvedt and Eva Jakobsson. Vol. 1. New York: I. B. Tauris.

Khan, I. 2009. "Clouded Details of Pakistan Deal." March 16. http://news.bbc.co.uk/2/south_asia/7946396.stm.

Laird, T. 2002. *Into Tibet: The CIA's First Atomic Spy and His Secret Expedition to Lhasa*. New York: Grove Press.

Lal, M., H. Harasawa, and D. Murdiyarso. 2001. "Asia." In *Climate Change 2001: Impacts*. Geneva, Switzerland: Intergovernmental Panel on Climate Change.

Lama, M. P. 2001. "Water Resources, Environment and Conflicts in South Asia." In *Intra and Inter-state Conflicts in South Asia*, edited by S. G. George. New Delhi: South Asian.

Mamdani, M. 1990. "Social Movements, Social Transformation and the Struggle for Democracy in Africa." *Council for the Development of Economic and Social Research in Africa Bulletin.*

Mandel, R. 1991. "Sources of International River Basin Disputes." Paper prepared for presentation at the annual meeting of the International Studies Association, Vancouver, British Columbia, Canada. Cited in Guest, "Conflict and Cooperation in a Context of Change."

———. 2004. *Security, Strategy, and the Quest for Bloodless War*. Boulder, CO: Rienner.

Matalas, N. 1997. "Reflections on Hydrology." In *Reflection on Hydrology*, edited by N. Burns. Washington, DC: American Geographical Union.

May, R. J., and B. Ray. 2006. *Corruption, Governance and Democracy in South Asia*. Kolkata, India: Towards Freedom.

Miller, F. 2006. "Environmental Risk in Water Resources Management in the Mekong Delta: A Multiscale Analysis." In *History of Water*, edited Terje Tvedt and Eva Jakobsson. Vol. 1. New York: I. B. Tauris.

Mukherji, P. N. 1986. "Social Conflicts and Social Change: Towards a Theoretical Orientation." In *Domestic Conflicts in South Asia*, edited by U. Phadnis, S. D. Muni, and K. Bahadur. Vol. 1. New Delhi: South Asian.

Mullick, H. A. M. 2009. "Obama's Afghanistan-Pakistan Quandary—Part II." Accessed April 22. http://yaleglobal.yale.edu/article.print?id=12256.

Nafey, A. 2008. "India and the G8: Reaching Out or Out of Reach." In *Emerging Power in Global Governance*, edited by A. F. Cooper and A. Antkiewicz. Waterloo, ON: Wilfrid Laurier University Press.

Nepali Times. 2007. "Stand by for the Great National Unravelling: Foreign Hand." February 9.

Nussbaum, M. 2007. *The Clash Within*. Cambridge, MA: Harvard University Press.

Osborne, M. 2004. *River at Risk: The Mekong and Water Politics of China and South East Asia*. Sydney: Lowy Institute.

Pacific Institute. 2007. "Bottled Water and Energy: Getting to 17 Million Barrels." Online_update@pacinst.org.

Phadnis, U., and R. Ganguly. 2001. *Ethnicity and Nation-Building in South Asia*. Rev. ed. New Delhi: Sage.

Ramprasad, C. 2007. "India's Middle-Class Failure." *PROSPECT*, September.

Rashid, A. 2009. "A Dangerous Void in Pakistan." Accessed March 9. http://yale global.yale.edu/display.article?id=12048.

Ray, B. 2007. *Internally Displaced Persons in India: A Ticking Time Bomb*. Kolkata, India: Alumni Association, Department of Economics, Calcutta University.

Renner, M. 1991. "Assessing the Military's War on the Environment." In *State of the World: A Worldwatch Institute Report on Progress Toward a Sustainable Society*. North Sydney: Allen & Unwin.

Reppeto, R. 1986. *Skimming the Water: Rent-Seeking and the Performance of Public Irrigation Systems*. Washington, DC: World Resources Institute.

Revkin, A. C. 2009. "What Next After Copenhagen?" Interview. December 18. http://www.abc.net.au/lateline/content/2009/s2776473.htm.

Rizvi, Hasan-Askari. 2000. *Military, State and Society in Pakistan*. New York: St. Martin Press.

Rogers, A. D. 2009. "How Not to Kill the Oceans for Fish." *Yale Global*, September 18.

Rowland, M. 2009. "Pakistan, Afghanistan Recommit to Fight against Taliban." Australian Broadcasting Commission. May 7. http://www.abc.net.au/lateline/content/2008/s2564083.htm

Roy, S. 2007. "We and Our Neighours." *Statesman Weekly*, May 19.

Shah, R. 1972. "Nepal: Reflection on Issues and Events of 1971." *Asian Survey* 12 (2): 116–20.

Shahzad, A. 2009a. "Bloody Urban Battles Could Lie Ahead in Pakistan." May 17. http://news.yahoo.com/s/ap/20090517/ap_on_re_as/as_pakistan/print.

———. 2009b. "Pakistan to Arm Village Militias to Fight Terror." Accessed February 23. http://news.yahoo.com/s/ap/20090222/ap_on_re_as/as_pakistan.

Sharma, S. D. 2003. "Indian Politics." In *Understanding Contemporary India*, edited by S. Ganguly and N. DeVotta. Boulder, CO: Rienner.

Shiklomanov, I. A. 1990. "Global Water Resources." *Nature and Resources* 26 (3).

———. 1993. "World Freshwater Resources." In Gleick, *Water in Crisis*.

Shiklomanov, I. A., and N. V. Penkova. 2003. "Methods of Assessing and Forecasting Global Water Use and Water Availability." In *World Water Resources at the Beginning of the 21st Century*, edited by I. A. Shiklomanov and J. C. Rodda. Paris: UNESCO.

Spector, L. S. 2009. "Pakistan, Taliban and Global Security—Part I." May 8. http://yaleglobal.yale.edu/display.article?id=12349.

Toosi, N. 2009. "Pakistan Truce Includes Enforcement of Islamic Law." February 16. http://news.yahoo.com/s/ap/20090216/ap_on_re_as/as_pakistan.

Toosi, N., and A. Shahzad. 2009. "Rising Tensions Threaten Pakistan, Taliban Peace." May 4. http://news.yahoo.com/s/ap/20090504/ap_on_re_as/as_pakistan/print.

United Nations. 1989. *E/ECE 1197, ECE/ENVWA/12: Charter on Ground-Water Management*. Geneva, Switzerland: Economic Commission for Europe.

United Nations Conference on Sustainable Development. 1997. *Comprehensive Assessment of the Freshwater Resources of the World: Report of the Secretary General*. Geneva, Switzerland: UN.

Weisman, S. R. 2004. "For 2 Ex-generals, a Common Language." Correction appended. *New York Times*, March 1.

Windrem, R. 2009. "Pakistan Expanding Its Nuclear Capability." MSNBC. May 13. http://yaleglobal.yale.edu/display.article?id=12364.

World Bank. 1991. *India Irrigation Sector Review*. Washington, DC: World Bank.

Zaidi, H. 2007a. "End Game." *India Today*, September 17.

Zaidi, S. Akbar. 2007b. "Musharraf and His Collaborators." *Economic and Political Weekly* 42 (45–46).

Zakaria, F. 1997. "The Rise of Illiberal Democracy." *Foreign Affairs* 76 (6).

Index

About the Author

Binayak Ray is a visiting fellow in the Department of Political and Social Change at the Australian National University. He is a former director of the Australian Agency for International Development (AusAID), the overseas aid agency of the Australian government.

www.ingramcontent.com/pod-product-compliance
Lightning Source LLC
LaVergne TN
LVHW101446200926
848483LV00001B/60
* 9 7 8 0 7 3 9 1 6 4 1 4 3 *